No. 1536
$14.95

BEGINNER'S GUIDE TO
READING
SCHEMATICS

BY ROBERT J. TRAISTER

TAB **TAB BOOKS Inc.**
BLUE RIDGE SUMMIT, PA. 17214

FIRST EDITION

FOURTH PRINTING

Printed in the United States of America

Library of Congress Cataloging in Publication Data

Traister, Robert J.
   Beginner's guide to reading schematics.

   Includes index.
   1. Electronics—Diagrams.   I. Title.
TK7866.T7   1983   621.381′022′3   82-19282
ISBN 0-8306-0136-8
ISBN 0-8306-1536-9 (pbk.)

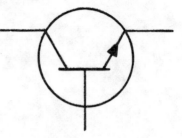

Contents

To my good friend, Jim Burn, from whom I learned much about electronic circuits through his schematic diagrams, which were (more often than not) drawn on a paper napkin over a cup of coffee in a local restaurant.

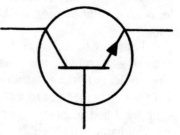

Introduction

Many people shy away from electronic pursuits because they think reading and drawing schematic diagrams will be complex and difficult. The fear of the unknown, however, is quickly erased when just a bit of knowledge is conveyed about this subject area. Refusing to enter an electronics pursuit because of schematic diagrams is equivalent to refusing to go swimming because of a fear of lifeguards. The lifeguard is put there to make swimming safer and easier. This is the prime purpose of schematic diagrams as well.

A schematic diagram is a road map of an electronic circuit; with it you have your own personal guide to understanding simple circuits, complex circuits, and even massive systems. Reading a schematic diagram is no more difficult than reading a road map once you have the proper background.

The purpose of this book is to provide the basic information you need to begin exploring electronic circuits and then to show you how to use that knowledge for circuit analysis, troubleshooting, and repair. This book explains in clear language the reason for schematic diagrams, how each symbol is derived, used and drawn, and how individual symbols are combined to make electronic circuits.

Before you can even dabble in electronic circuits, it is mandatory that you have a working knowledge of schematic diagrams and their uses. I taught myself to read schematic diagrams at the age of eight; many of my experiences, problems, and solutions form a basis for the discussions in this book.

Learning to read schematic diagrams and to use them to analyze electronic circuits is one of the simplest parts of electronic experimentation, design, and troubleshooting. They are provided not to make the field more difficult, but to simplify it and adapt it to human beings. Schematic diagrams are "human-engineered" to allow the interfacing of human deductive powers with electronic circuits.

This book will take you through the basics and provide enough information to allow you to continue to perfect the ability to delve deeply into electronic circuits with an understanding of function and design. This is the first step on your journey to electronics proficiency.

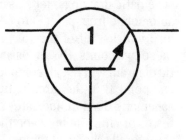

What Are Schematic Drawings?

A schematic diagram is a map of an electronic circuit showing each and every component and how they are interconnected. According to *Webster's, schematic* means "of or relating to a scheme; diagrammatic." Therefore, any diagram which depicts a scheme—be it electronic, electrical, physiological, or whatever—can be classified as a schematic drawing. Most of us depend upon schematic drawings every day of our lives to give us the whole picture of a scheme or event within a confined medium. This aids the understanding of the scheme as viewed in its entirety.

One of the most common schematic diagrams is used by nearly everyone who has ever driven an automobile. It's called a *road map* and is a diagrammatic form of representing an entire scheme. The scheme may involve the paths of travel within a small locality, within a state or even several within states. Like a schematic diagram of an electronic circuit, the road map shows all the components which are relative to the particular travel scheme it addresses. Motorists will make up their own schemes which are often small portions of the total scheme included in the road map. Likewise, electronic schematic diagrams may be used to show the whole scheme but also to allow the technician to extrapolate the section which fits the scheme he has in mind.

Using the road map as a comparison to electronic schematic drawings, let's assume you wish to travel in your automobile from point A to point B. We can say that the road map will list all of the towns and cities which lie between these two points. By comparison, it can be said that a schematic diagram will list all components between a similar point A and point B in the circuit. But this is not all either type of schematic diagram indicates. It's not enough to know which towns or cities lie between these two points to get an idea of the overall scheme of things. Indeed, we could far more easily write down the names of these locations in which case we would not have to resort to a diagram at all. From the electronics standpoint, we could do the same thing by simply providing a list of the components that were used to build a certain circuit such as:

- ☐ 120 ohm resistor
- ☐ 1000 ohm resistor
- ☐ pnp transistor
- ☐ .47 uf capacitor
- ☐ 2 feet of hookup wire
- ☐ 1.5 volt battery
- ☐ switch

Now, what has this list told us about the circuit? Not much, really. We do know the components that are involved in building it, but unfortunately, we don't know what it is that was built.

Any schematic drawing must not only indicate all components necessary to make up a specific scheme but also *how* these components are interrelated . . . how they are connected. The road map interconnects the various towns, cities, and other trip components by lines which represent streets and highways. A line which indicates a secondary road is different from one which is used to represent a four-lane highway. With a bit of practice, we can tell which lines indicate which roads. Likewise, an electronic schematic drawing uses lines to indicate a standard conductor; other types will be used to represent a cable. In both cases, when the interconnecting lines are added, a relationship is established between the connected components. Then too, the physical relationship of one component of a road map to others tells us something as well. This is not as true in electronic diagrams but does provide a type of relationship which is enforced by interconnecting lines representing conductors.

Symbols

Now it becomes necessary to talk about *how* a schematic diagram displays the scheme of a system. It does this through symbology. The lines which form indications of roadways are symbols. Symbols are used here as opposed to pictorial drawings. A single black line which may indicate Route 522 in no way resembles the actual appearance of this highway. It is enough for us to know that a black line is Route 522. We can make up the rest in our own minds. If it were necessary to provide a pictorial drawing of this route, road maps would be a hundred times larger in size and probably ten thousand times more difficult to read.

The same is true of towns, cities, railroad tracks, airports, and a thousand other features found on a standard road map. You cannot pictorially represent them in a practical manner. Instead, these components are represented by symbols. A key to the symbols used is often provided on the map. It shows the symbol and explains in plain language what each means. If a small airplane drawn on the map indicates an airport and this fact is known to us, then each time we encounter the airplane symbol, we will know that an airport exists at this site. Again, we are able to visualize the airport in our minds using the tiny drawing as a stimulus. Symbology involves the depiction of a physical object (in this case only) by means of another physical object (the miniature airplane) for the purpose of practical representation.

The airplane drawing could be replaced with a drawing of a beer bottle and it would mean the same thing as long as the miniature bottle of suds was correctly identified in plain language as representing an airport. In individual cases, the makeup of the symbol is not that important.

However, a road map must contain many different symbols. Each is usually human engineered to be logical to the human mind. For instance, when you see a miniature airplane on a road map, you will be inclined to think that this area had something to do with airplanes, so a detailed explanation would not be necessary. If, on the other hand, you used the ridiculous example of a beer bottle to represent the same thing, anyone who did not read the key would certainly not be inclined to think of an airport. Since many different symbols would be used, it is mandatory that whenever possible, each be presented in a logical manner.

Logic will only take it so far, however. This is especially true in electronics diagramming, where the actual appearance of a component may vary greatly from manufacturer to manufacturer. A certain logic does apply, however, after a basic groundwork has been laid in electronics symbology. For example, a circle is used to indicate a vacuum tube. Other symbols are used inside this circle to represent the many tube electrodes. A tube is an active device, capable of producing an output which is of higher amplitude than the signal at its input. The same can be said of a transistor, which was developed many years after vacuum tubes. Since a circle with electrode symbols had been used for many years to represent vacuum tubes and due to the fact that transistors were developed as active devices to take the place of some tubes, the schematic symbol for the transistor also started with a circle. Electrode symbols were inserted into this circle as before, but the symbols here were different from tube elements, so the two types of devices could be easily distinguished. The logic here is based around the circle symbol. Transistors accomplish many of the same functions in electronic circuits as vacuum tubes do, so symbolically they are quite similar as to circuit function. It would not be logical to develop a symbol that was far removed from the vacuum tube symbol, which had been used for decades.

There are certain inconsistencies, however. Circles will sometimes make up a part of an electrical symbol indicating solid-state devices which are *not* symbolically equivalent to tubes and transistors. A zener diode, for example, is often indicated by a circle with a special diode symbol at its center. A zener diode is not a transistor, and the electrode symbol at the center clearly indicates that this is not a transistor. Fortu-

4

nately, the circle symbols today are most often used to indicate either a vacuum tube or some sort of solid-state device. This is not always the case, but it is often so. Chapter 2 will describe the various components which are represented schematically and will explain how each symbol is to appear and how it is drawn.

Interconnections

To explain further how schematic diagrams are used, we can take a single component, a pnp transistor. This device has three electrode elements, and while there are many thousands of different varieties of pnp transistors, they will all be drawn symbolically in the same manner. You must remember that there are thousands of different circuits into which a single pnp transistor can be inserted. What schematics are really needed for is to indicate *how* the transistor is connected in the circuit, what other components are used in conjunction with this device, and what other circuit portions depend upon this device for overall operation. A transistor, for example, may be used as a solid-state switch, an amplifier, and impedance matching device, etc. One type of transistor may serve all of these purposes. Therefore, if a transistor is used in one circuit as an amplifier, you cannot say that this transistor is used as an amplifier only. As a matter of fact, you could pull this transistor out of the amplifier circuit and put it into another one to form a solid-state switch. By knowing the type of component alone, you cannot know how it is used in a circuit until you can get an overall view of all connections and interconnections. This can't be done, in most instances, by examining the physical circuit. You need a road map, a schematic diagram, to show you all the connections that were made to form the circuit.

A schematic is needed because the human brain cannot retain in memory all of the input data which is fed to it by the eyes when scanning a small portion of a physical circuit. As a practical non-electronic example, let's assume that you are to drive from Washington, D.C. to Los Angeles, California. Even if you had made the trip several times before, there's a good chance that you could not remember all of the routes to take and all of the towns and cities that you passed on the way. A road map, however, would give you an overall picture of the entire trip. Since all of the trip data has been collected and presented in a form which can be scanned at a glance, the road map becomes instrumental in allowing you to see the entire trip rather than a piece of it at a time. The schematic diagram does the same thing for a trip through an electronic circuit.

Let's talk about troubleshooting and electronic circuit repair. Using the road map and the coast-to-coast trip again as an example, let's assume that you have memorized the entire route. Assume also that one of your prime routes of travel along the way is under construction, and it becomes necessary to take an alternate route. Without a road map, you would not know a separate artery to take, one which would keep you on course as much as possible and eventually return you to the original travel path.

In an electronic circuit, there are many electrical highways and byways. Occasionally, some of these break down, making it necessary to seek out the problem and correct it. Even if you can visualize the circuit in your head as it appears in physical existence, it is nearly impossible to have any idea of the many different routes which are used, one or more of which may be defective. When I speak here of visualizing the circuit, I am not speaking of the schematic equivalent of the circuit, but the actual hard wiring itself. A schematic diagram is necessary to give you an overall picture of the circuit and to show how the various routes and components depend upon other routes and components. When you can see how the overall circuit depends upon each individual circuit leg and component, it then becomes easier to diagnose the problem and to effect repairs.

Symbology

It is often difficult to fully expain schematic diagramming to individuals who are just starting out in electronics. One must think of this form of symbology as a language. What is a language? It is a system of symbols which are used to communicate ideas. The English language is a symbology with which most of us are quite familiar.

6

Every word spoken in English or any other verbal language is a complex symbol made from simple symbols called numbers and letters. Let's take the word "stop," for example. Without a reference key, it means nothing. However, through learning the symbology from shortly after birth, this word begins to mean something, because the infant who is learning to speak and understand can compare "stop" to other words, but especially to actions. I realize that it may sound like this discussion is getting very deep, but this is necessary for a true understanding of schematic symbology. We can even say the word "stop" is symbology within symbology. The communication's intent of this word can also be expressed by the phrase "Do not proceed further." This phrase, however, is still symbology, expressing a mental image of a desired action.

If we could all communicate by telepathy, then symbology would not be necessary. There are those who would argue with this, saying that we all think in our own language, language being symbology. I do not believe this to be true. Thinking is done on a far speedier level and is identical from human to human, regardless of what language or languages he understands. A newborn baby, for instance, speaks and understands no language whatsoever. However, whether that baby was born in the United States, South Africa, Asia, or wherever, thought processes do take place. The baby knows when it is hungry, in pain, frightened, etc. It needs no language to comprehend this. It does become necessary to communicate right from the start. For this reason, all newborns communicate in the same language (crying, mostly). As newborns are able to comprehend more and more of their environment through improved sensory equipment (eyes, ears, etc.) more data is collected. The environment then plays a role in allowing the baby to comprehend communications symbology. This is where the various languages come into play, with different societies using different verbal symbols to express simple and complex mental processes. The human brain still carries on the same non-linguistic thought processes as before, because to think in terms of symbols would take far too much time and memory storage area. The brain does, however, allow the human being to transpose complex thoughts into a language. When a child is about to step in front of a speeding automobile, if the brain had to handle the words "automobile, speed, death, child" and literally millions of other criteria symbolically, we human beings would spend all of our lives waiting for it to deliver the correct processed information. Rather, the brain scans all that is received by the sensory organs in real time and then sums it up into a single symbol which can be used for communication. The symbol is the word "Stop!" When that word is communicated to the child a similar process takes place in its brain, having been triggered by the uttered symbol.

All languages do not involve the spoken word, however. We've all heard of Indian sign language, whereby the arms and hands are used to communicate ideas. In most instances, an entire communicating language made up of visual symbols only is not as efficient (to human beings) as one composed of words and visual symbols. This is partially due to the way human beings speak words. Using the symbol "stop" again, we know that this can be uttered in many different ways. The word in itself means something, but the way it is spoken augments the meaning. The tone of voice, inflection, and certainly the volume at which it is uttered can modify the basic meaning. It is sometimes difficult to do the equivalent by visual means only. We have arrived at some universal methods of modifying visual symbols. To many of us, the color red denotes danger or at least something that should be given immediate attention. Often, however, this is done in conjunction with the visual symbol for a spoken word.

As far as the symbology of schematic drawing is concerned, this is a visual language due to the makeup of human beings. It does not lend itself readily to any form of oral symbology. Our senses along with our central processor, the brain, render us less than proficient at mentally conceiving all of the workings of electronic circuits by dealing with them directly. Therefore, it is necessary to accept data a small step at a time, compiling it in hard copy form (through symbology) and providing a hard copy readout which is the schematic diagram. In order for us to understand what takes place in the entire circuit, all of this collected information must be presented at one time and in a form whereby we can see all of the steps which went into its making at one time. This is how schematic diagrams are prepared. This can be likened to the "connect-the-dots" drawings that are offered in many childrens' workbooks. Individually, the dots mean nothing. But once they are arranged in logical form and connected by lines, we get an overall picture. One cannot, however, deny that that picture was formed starting with a few dots. The dots themselves are unimportant, but their relationship to each other and to the order in which they are connected means everything.

Don't let this discussion on symbology lead you to think that schematic diagramming is a highly complex field. It is not; schematic drawings are provided to make things simple. We use symbology every day and take it for granted. Therefore, schematic diagrams are something we can logically take to . . . and in a short period of time.

The following chapters in this book will start with the basics of schematic diagramming, the symbols, and take you through to simple circuits and finally on to complex ones. Schematic symbols and diagrams are designed for humans; therefore, human logic is a prime factor in determining which symbols will mean what. Anyone who is able to read and do simple math can teach himself to read and draw schematic diagrams.

Schematic Symbols

On a true road map, different figures are used to illustrate towns, cities, secondary roads, primary roads, airports, railroad tracks, etc. The same applies to schematic drawings; different symbols are used to indicate conductors, resistors, capacitors, solid-state components, and other electronic parts. Every electronic component manufactured today has an equivalent schematic symbol. As new types of components come out which are completely different from all previous ones, a new schematic symbol is derived for each. This doesn't happen very often, even in these days of rapid advancement in the electronics field. Often, a new type of component is a modification of one which already exists. Therefore the schematic symbol is also a slight modification of the one used to indicate the preexisting component.

Resistors

Resistors are electronic components. They are aptly named, as they resist the flow of electrical current. The value of resistors is measured in *ohms* and typical components may be rated at less than one ohm or more than several million ohms. However, regardless of the resistance value, all resistors are schematically indicated by the symbol shown in Fig. 2-1. Some resistors have the ability to change value. These

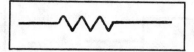

Fig. 2-1. Standard schematic symbol for a resistor.

will be indicated by a slight modification of the symbol shown. These will be dealt with later on. The resistor symbol is composed of three full upside down triangles and a half triangle on each end connected to lines which are horizontal, as in the example shown. This is the most universally accepted symbol for the resistor.

You may also see resistors drawn in other forms, as shown in Fig. 2-2. These symbols get the point across, but they do not strictly follow the standard format. Do not draw resistors in this manner when you are making up your own schematic diagrams.

The two horizontal lines are actually indicators of the leads or conductors which exit from both sides of the resistor. Sometimes the resistor contacts are not wire leads but are in fact metal terminals. Figure 2-3 shows a pictorial drawing of a carbon resistor with leads on either side. This is basically what the component will look like when you purchase it. Figure 2-4 shows pictorial drawings of several other types of resistors which do not contain the basic end leads. However, all resistors shown pictorially here will be indicated schematically by the basic resistor symbol of Fig. 2-1.

Variable resistors are those which have the ability to change resistance through a slide tap or a vernier control method. Schematic symbols are used to indicate a specific device but also serve as an indication of a specific function. The variable resistor is usually set to one value and then remains at this point until manually changed. The electronic circuit still sees this component as a lumped resistance. However, when a variable resistor is required for the proper functioning of a specific circuit, it is necessary to indicate to the person who may be building this circuit from a schematic

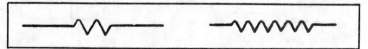

Fig. 2-2. Alternate resistor symbols.

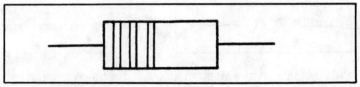

Fig. 2-3. A carbon resistor.

drawing that this particular component must be a variable type.

The schematic symbol shown in Fig. 2-5 is that of a *variable resistor*. Again, this is the most standardly recognized symbol. However, others have crept into schematic drawings for many years and may look more like the example shown in Fig. 2-6. Notice that both examples use the standard resistor configuration but indicate that it is a variable type by using an arrow symbol in conjunction with the triangles. In schematic drawings, an arrow is often used to indicate variable properties of a component—but not always, so don't assume too much at this point. Most types of transistors, diodes, and other solid-state devices also use an arrow as a part of their schematic symbols. This in no way indicates any variable properties. A true variable resistor has only two contact points or leads, as indicated by the schematic drawing. Figure 2-7 shows a pictorial example of a variable resistor. These are most often wire-wound units which have been manufactured so that some of the resistance wire is exposed on the surface of the component. A sliding metallic collar is wound around the body of the resistor and may be adjusted to intercept a different turn of resistance wire. This collar is also attached by a flexible conductor to one of the two resistor leads. The collar effectively shorts out resistance turns, so as it is advanced toward the opposite resistor lead, the ohmic value of the component decreases.

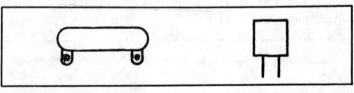

Fig. 2-4. Other types of resistors.

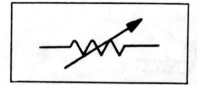

Fig. 2-5. Schematic symbol for a variable resistor.

Figure 2-8 shows the schematic drawing for a *rheostat* or *variable resistance control.* Notice that this symbol looks very much like the variable resistor equivalent but has three discrete contact points. Using the rheostat control, the portion of the circuit which comes off the arrow lead can be varied in resistance to two circuit points, each connected to the two remaining control leads. Figure 2-9 shows a drawing of such a control.

Rheostats and most variable resistors are closely related. As a matter of fact, the variable resistor discussed earlier can be changed into a rheostat by simply severing the lead between the collar and one end lead. Now, the collar can be used as the third or variable contact and a rheostat is formed. Likewise, a rheostat can be turned into a two-lead variable resistor by simply shorting out the variable lead with one on either end. Rheostats are often used as variable resistors in this manner.

Resistors of all types often make up the most numerous elements in many electronic circuits. It has already been stated that resistors may carry a component value ranging from a fraction of an ohm to several million ohms. The schematic symbol proper in no way gives any indication of the value of the resistor. It is used to indicate a resistor is in the circuit. The actual value of the component may be written alongside the schematic drawing, but it may also be given in a separate components table which is referenced by an al-

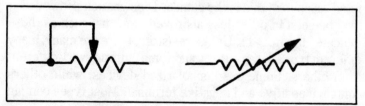

Fig. 2-6. Alternate variable resistor symbols.

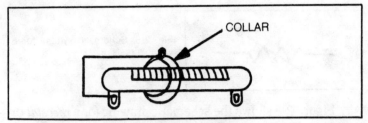

Fig. 2-7. Pictorial drawing of a variable resistor.

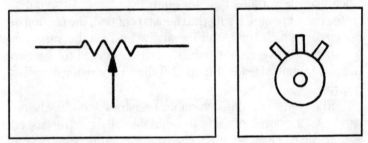

Fig. 2-8. Schematic diagram of a rheostat.   Fig. 2-9. A rheostat control.

phabetic/numeric designation printed next to the schematic symbol. This will be discussed in a later chapter.

Capacitors

Capacitors are electronic components which have the ability to block direct current while passing alternating current. They also can be used to store power. The basic unit of capacitance is the *farad*. This is a tremendously large quantity, and most practical components will be rated in microfarads or in picofarads. A *microfarad* is equivalent to 1/1,000,000th of a farad, while a *picofarad* is 1/1,000,000th of a microfarad. Next to resistors, capacitors make up the largest types of components in many electronic circuits.

Figure 2-10 shows the schematic symbol of the basic fixed capacitor. This is the standard designation. However, it may be seen in other less approved of forms, such as those indicated in Fig. 2-11. Unlike resistors, there are many, many different types of capacitors, and some units are very different from others. Some are nonpolarized devices, while others contain a positive and negative terminal. Most types contain only two leads.

14

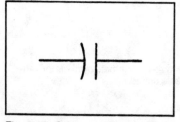

Fig. 2-10. Schematic symbol for a fixed capacitor.

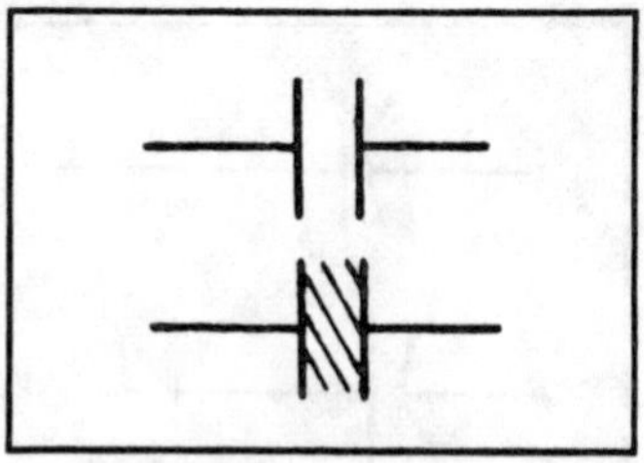

Fig. 2-11. Alternate symbols for fixed capacitors.

The basic capacitor symbol consists of a vertical line followed by a space and then a half moon symbol. Horizontal lines connect to the centers of the vertical line in the half moon to indicate the component leads. The symbol discussed indicates a *nonpolarized* capacitor which may be made from ceramic, mica, mylar, or some other material. The material designation here indicate the insulation which is used between the two major parts of the component. As the symbol might indicate, a capacitor is simply two tiny sheets of conductive material which have been placed in close proximity to each other.

Figure 2-12 shows the schematic symbol for a *polarized* or *electrolytic* capacitor. Notice that this is nearly identical to the former symbol, but a plus (+) sign has been added to the vertical line side. This indicates that the positive terminal of the component will be connected to the remainder of the circuit, as indicated in the schematic drawing. Occasionally, a negative (−) symbol will also be drawn on the opposite side (over the half moon lead), but this does not follow standard schematic form. When the positive symbol is seen, this identifies the component as an electrolytic capacitor which must be connected to the remainder of the circuit in observance of correct polarity, i.e., the positive capacitor elec-

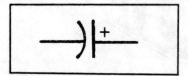

Fig. 2-12. Schematic symbol for an electrolytic capacitor.

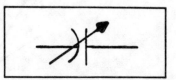

Fig. 2-13. Standard symbol for a variable capacitor.

15

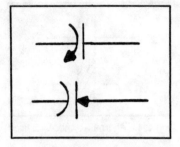

Fig. 2-14. Alternate symbols for variable capacitors.

trode must be connected to the positive voltage lead of the remaining circuit. Electrolytic capacitors themselves contain case markings which indicate the positive lead.

To this point, all capacitors and their symbols have been of fixed design. In other words, the components specified do not have the provisions for changing their capacitance value, which is fixed at the time of manufacture. Some capacitors do have the ability to change value. These are usually called *variable capacitors*, but some specialized types may be known as *trimmers* and/or *padders*. Figure 2-13 shows the basic symbol for a variable capacitor. Again, an arrow is used to indicate the variable property and is drawn horizontally and through the fixed capacitor symbol. Figure 2-14 shows other ways of indicating this same component, although these are rarely used. Most of the time, the standard symbol shown in Fig. 2-13 will indicate a variable capacitance, regardless of the exact construction of the component. Figure 2-15 shows

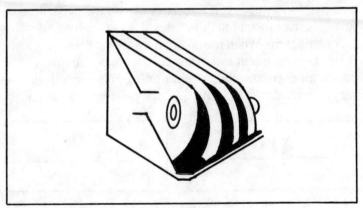

Fig. 2-15. An air-variable capacitor.

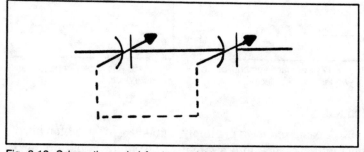

Fig. 2-16. Schematic symbol for two variable capacitors which are ganged together.

an air variable capacitor of the type which is used to tune many AM radios. This component still consists of two basic plates, but their proximity to each other can be changed. In the example shown, there are many interlaced plates, but every other one is connected to the first, forming two distinct contact points.

Variable capacitors are always nonpolarized. The electrolytic capacitor is the only type which will carry a polarity designation. Many circuits will quite frequently contain a fair sampling of all three types of capacitors discussed here. Sometimes, two separate variable capacitors will be connected together or ganged in a circuit. This means that two or more units are used to control two or more electronic circuits, but both components are varied simultaneously. This is accomplished by tying the rotors of the two units together. The rotating plates in a capacitor are referred to as the *rotors*, while the fixed plates are designated the *stators*. Figure 2-16 shows the schematic symbol for two variable capacitors which are ganged together. This is a simple alteration of the basic variable capacitor symbols and involves drawing a dotted line beneath each and connecting it with a horizontal dotted ine.

As is the case with most electronic components, the schematic symbol for the capacitor only serves to identify it as much and to tell whether it is fixed, variable or polarized (electrolytic). The component value may be written alongside, or it may be given a letter and number designation for reference to a components list.

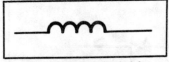

Fig. 2-17. Standard symbol for an air-wound coil or inductor.

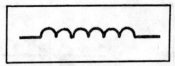

Fig. 2-18. Alternate symbol for air-wound inductor.

Inductors

A basic *inductor* is simply a coiled wire used for introducing inductance into a circuit. *Inductance* is the property which opposes change in existing current and is present only when the current is actually changing. *Coils* or inductors may be tiny or very large, depending upon the inductance value of the component. The basic unit of inductance is the *Henry*, although it is a very large electrical quantity. Therefore, most practical inductors are rated in *millihenrys* (1/1,000th of a Henry) or *microhenrys* (1/1,000,00th of a Henry).

Figure 2-17 shows the basic schematic symbol for an *air-wound* coil. It consists of a single line which has been used to form five loops. Two leads are designated by the straight lines which eventually curve into the coil at the top and bottom of the symbol. An air-wound coil is one which specifically has no internal core between the windings. However, in practice, a nonconductive and noninductive form such as plastic, mica, or some other insulator will be used as a support for the turns. This is still an air-wound coil, however. While the symbol shown for the basic inductor is the most approved one, you will often encounter the same component drawn as in Fig. 2-18. This one is as generally accepted and is quite a bit easier to draw when making up your own schematics. Both, however, indicate air-wound, fixed-value inductors.

Figure 2-19 shows the schematic symbol for a type of *variable* air-wound inductor. This is more accurately known as a *tapped coil* or *tapping arrangement*. Whereas the fixed coil had only two leads, the tapped coil may have three or more. When a coil is tapped, separate conductors are attached to one or more of the turns to offer an alternate connection point. Maximum inductance is obtained from connecting the coils to the circuit at either end. A tapped arrangement allows for selection of an input or output point which offers lower induc-

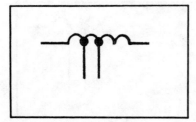

Fig. 2-19. Schematic symbol for a tapped inductor.

tance. Some coils are fitted with a sliding contact which can be continuously advanced throughout the entire coil. This allows for continuous adjustment of the inductance value rather than selecting a fixed point as was done with the tapping arrangement. A *continuously variable* coil is often indicated by the symbols shown in Fig. 2-20. This indicates the component is capable of being continuously adjusted from a maximum inductance value (determined by the physical size of the coil) to minimum value. Figure 2-21 shows an example of a fixed, tapped, and continuously adjustable air-wound coil.

Not all coils are of the air-wound variety. *Chokes* and other types of coils which are used for low-frequency applications may consist of a coiled conductor which has been wound around an *iron core.* Here, the iron material replaces the previous empty or air core. 60 Hertz chokes, for example, may closely resemble an ac power transformer (discussed later) and will contain a single coil wound around a circular iron form. Figure 2-22 shows the schematic symbol for an iron core inductor. Notice that this is the basic fixed coil discussed earlier, which is immeditaely followed by two close-spaced vertical lines which run for the entire length. Sometimes you will encounter the iron core inductor drawn as shown in Fig. 2-23. Here, the vertical lines are placed inside the coil turns in the symbol. This is not the approved method of indicating an

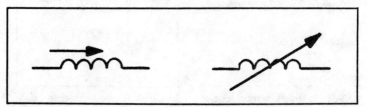

Fig. 2-20. Schematic symbol for a continuously variable inductor.

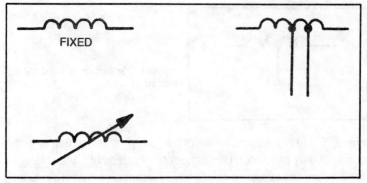

Fig. 2-21. Fixed, tapped and adjustable air-wound coils.

iron core inductor. However, it is encountered quite frequently. Some iron core inductors may also contain taps for sampling different inductance values and some may even be continuously adjustable. The equivalent schematic symbols for these types of components are shown in Fig. 2-24.

At higher frequencies, iron cores are far too inefficient to be used in inductors. Especially at rf frequencies, a special core is needed and is most often composed of iron material which has been shattered into many tiny fragments, each of which is insulated. After this process has been completed, the particles are greatly compresssd to form what appears to be a solid core. Here, the material is called *ferrous iron* or *ferrite* and is known as a *powdered iron core.* Figure 2-25 shows the schematic symbol for a powdered iron core inductor. It can be

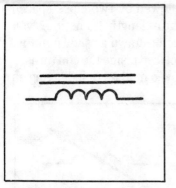

Fig. 2-22. Schematic symbol for an iron-core inductor.

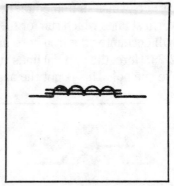

Fig. 2-23. Alternate iron-core inductor symbol.

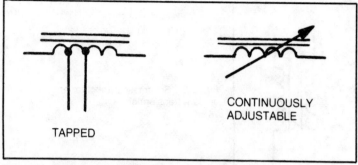

Fig. 2-24. Schematic symbols for adjustable iron-core coils.

seen that the symbol is nearly identical to the iron core inductor, with the exception that the two vertical lines are broken at several different points. These types of components may also be tapped or made continuously variable. The schematic symbol will be identical to those discussed earlier for variable iron core inductors, except that the broken vertical lines will remain.

Transformers

Transformers are closely related to inductors and are made when the turns of two or more coils are interspersed. Figure 2-26 shows a basic air-core transformer which consists of two air-core coils drawn back to back. A transformer is a device with the ability to transform electric energy from one circuit to another at the same frequency. Since transformers are made by combining inductors, the schematic symbols are very similar. Figure 2-27 shows other types of transformers which contain iron cores, powdered iron cores, are variable,

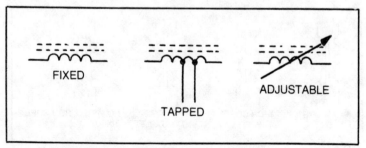

Fig. 2-25. Schematic symbol for a powered iron-core inductor.

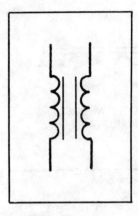

Fig. 2-26. Schematic symbol for an air-core transformer.

tapped, etc. Notice that many of the additional symbol lines and indicators are identical to those used in inductor symbology.

Switches

A *switch* is a device, mechanical or electrical, that completes or breaks the path of current. Additionally, a switch may be used to allow current to pass through different circuit elements. Figure 2-28 shows the schematic symbol for a *single pole single throw* (spst) switch. The spst variety is capable of making or breaking a contact at only one point in a circuit. Notice that the arrow is used here to indicate variability. This type of switch is a two-position device (on-off, make-break). Figure 2-29 shows a different type of switch which is designated as a *single pole double throw* (spdt) variety. For basic understanding, the pole is the point of contact at the base of the arrow. The throw(s) is the contact point to which

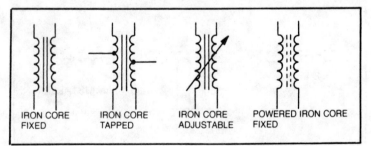

Fig. 2-27. Schematic symbols for other types of transformers.

22

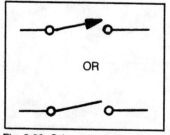

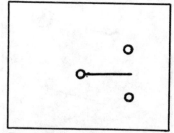

Fig. 2-28. Schematic symbol for an spst switch.

Fig. 2-29. Schematic symbol for an spdt switch.

the arrow can be attached. The spdt switch contains one pole contact and two throw positions. This means that the input to the pole may be switched to either the left-hand or right-hand circuit point.

Some switches contain two or more poles. An example of a *double pole single throw* (dpst) switch and a *double pole double throw* (dpdt) switch is shown in Fig. 2-30. Some switches may have even more elements. The one shown in Fig. 2-31 has five poles, each of which can be switched to two separate positions. Therefore, this is a *five pole double throw* arrangement and is designated as 5 pdt.

This last designation can actually be covered under the heading of *multi-contact* switches. This category takes in most switches which have more than two poles or two throw positions. For example, a *rotary* switch is a device which has a single pole and sometimes ten or more throw positions. This

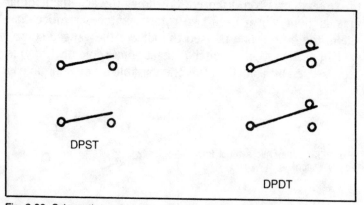

Fig. 2-30. Schematic symbols for a dpst and a dpdt switch.

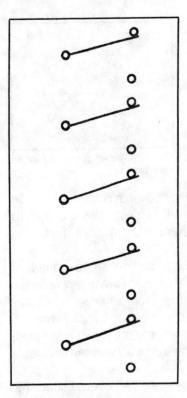

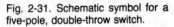

Fig. 2-31. Schematic symbol for a five-pole, double-throw switch.

basic category of switch is shown in Fig. 2-32. The arrow still indicates the pole contact.

Occasionally, rotary switches will be ganged together, much like variable capacitors were ganged together in a previous discussion. Figure 2-33 shows the schematic of an arrangement which uses two rotary switches. Notice that again, the dotted line is used to indicate the ganged setup.

In each case, the switch contact point (pole or throw) is represented by a small circle. The variable element or pole is

Fig. 2-32. Schematic symbol for a rotary switch.

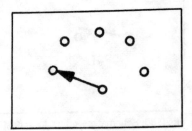

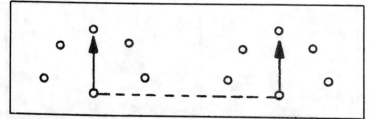

Fig. 2-33. Schematic symbol for two rotary switches ganged together.

indicated by an arrow. The symbols shown are all standard. Due to the simplicity of this symbol, variations are seldom seen. There is a special type of switch, however, which is used in some radio communications work. This is called a *code key, Morse key,* or simply *key*. This is the device used to make and break an electrical contact for the purposes of sending code. A key is simply an spst switch which contains a spring that allows it to return to the off position automatically. The symbol for a key is shown in Fig. 2-34.

Conductors and Cables

Throughout this discussion, a straight line has always been used to indicate a conductor. This is easy to understand, but most circuits contain a large number of conductors. It often becomes necessary to have them cross over each other or to actually make contact. Figures 2-35 and 2-36 show the standard procedure for indicating what is occurring when two conductors cross. Figure 2-35 shows two conductors which have crossed but not made contact. This does *not* mean, by any stretch of the imagination, that when building the circuit, the conductors must cross over each other. It simply means that in order to make the schematic drawing, it became necessary to draw one conductor across another to reach various circuit points. Notice that at the junction, a half loop is drawn

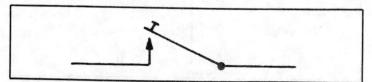

Fig. 2-34. Schematic symbol for a code key.

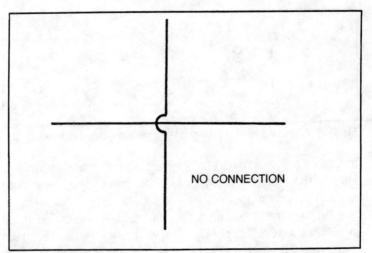

Fig. 2-35. Schematic symbol indicating conductors which cross over.

in one conductor to indicate that no connection has occurred. The loop could just as easily have been drawn in the horizontal line to indicate the same condition.

Figure 2-36 shows an actual wiring connection. Here, the two conductors cross (at right angles in this example), and a black dot is drawn at the junction. This indicates the conductors connect at this point. The drawing of conductors is the

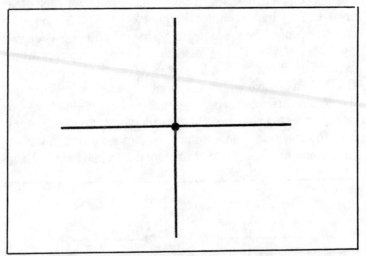

Fig. 2-36. Schematic symbol indicating conductors which are connected.

most often abused part of schematic preparation. Occasionally, a non-connection will be drawn as shown in Fig. 2-37. A connection will still be indicated by a black dot at the junction, but it still leaves you wondering. Sometimes this same arrangement will be used to indicate a connection, and a no-connect crossover will be indicated by the half loop. When other than standard methods are used, it becomes necessary for the person reading the schematic to deciper just what the artist means by hunting down a non-connect example or a connect example to compare the other to.

A *cable* consists of two or more conductors usually contained in the same insulating jacket. Quite often, unshielded cables will not be specifically indicated in a schematic drawing but are simply shown as two discrete symbol changes. Figure 2-38 shows an example of a *shielded wire*. This is often used to indicate the use of *coaxial cable* in an electronic circuit, especially one which is operating at radio frequencies. It consists of a single conductor surrounded by a metallic shield for the entire length of the cable. An insulator keeps the two conductive elements isolated from each other. The symbol for

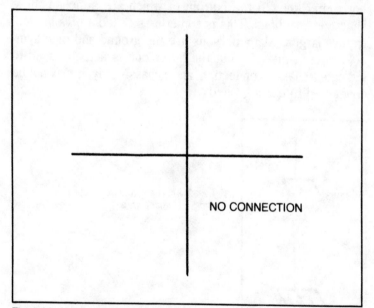

NO CONNECTION

Fig. 2-37. Alternate method of indicating a crossover without connection.

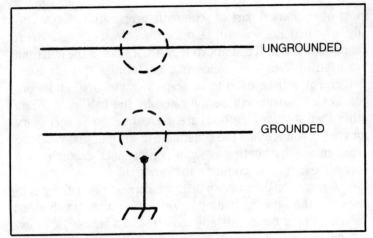

UNGROUNDED

GROUNDED

Fig. 2-38. Schematic symbol for a shielded wire.

the shielded wire is drawn by placing a small circle over the conductor. The circle is attached to a vertical conductor which, in turn, is connected to ground. Here, the ground symbol consists of four tapered lines. Alternately, the ground symbol may be indicated as shown in Fig. 2-39. This is a horizontal line, to the bottom of which are attached three short vertical lines. This is sometimes called a *rake* in electronics jargon. Many persons use the ground and rake symbols interchangeably, but the latter one is actually used to indicate a chassis connection, and a chassis may or may not be connected to earth ground.

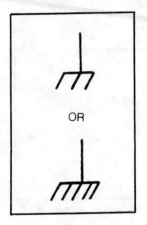

OR

Fig. 2-39. Alternate method of representing a shielded wire.

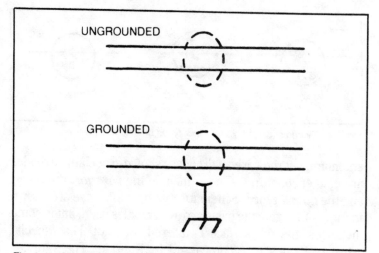

UNGROUNDED

GROUNDED

Fig. 2-40. Schematic symbol for a shielded cable.

Sometimes a shielded cable is necessary for electronic construction. This consists of two or more conductors which are surrounded by a single shield. The schematic symbol for this is shown in Fig. 2-40. This is identical to the single shielded wire symbol, with the exception that an extra conductor has been added. If the cable were to contain five conductors, then the circle (indicating the shield) will contain five horizontal lines.

Solid-State Components

Solid-state components are numerous and varied and include *transistors, diodes, thyristors, solar cells*, and other specialized devices. Figure 2-41 shows the basic symbol for a diode or rectifier. It is composed of an arrow lead contacting a flat surface which contains another lead. Several ways are shown for indicating a diode, but the most prevalent one is indicated first. Diodes may be made of germanium, silicon, or

Fig. 2-41. Schematic symbol for a diode or rectifier.

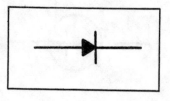

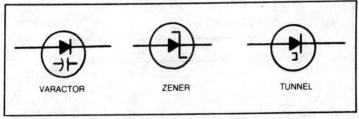

| VARACTOR | ZENER | TUNNEL |

Fig. 2-42. Schematic symbols for other types of solid-state diodes.

selenium, but they are all drawn as shown, Specialized diodes are also encountered. These include the *varactor*, the *zener*, and the *tunnel* types. Schematic symbols for these are shown in Fig. 2-42. Again the basic arrow symbol is used, but in some instances, the flat surface is altered. Each of these special types of diodes is encompassed by a circle. The varactor diode contains a tiny capacitor symbol inside the circle. This diode has the ability to change capacitance, which is the reason for this addition. Often, zener and tunnel diodes will be drawn without the containing circle, and sometimes they will be indicated as a standard diode but are further defined in the components list.

A *silicon-controlled rectifier* (SCR) is a special three-element diode, and the schematic symbol is shown in Fig. 2-43. Again, the circle is used and the third element (the gate) is indicated by a diagonal line connected to the component lead symbol. In all cases, the lead which is attached to the arrow is the *anode* of the device, whereas the one connected to the flat surface is the *cathode*.

Figure 2-44 shows the basic schematic symbols for *bipolar transistors*. The *pnp* type is shown first, followed by the *npn* variety. The only distinction between the two is the

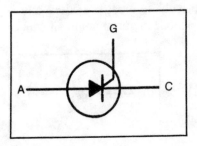

Fig. 2-43. Silicon-controlled rectifier schematic symbol.

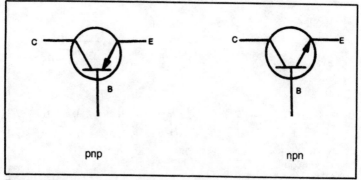

Fig. 2-44. Basic symbols for bipolar transistors.

direction the arrow is pointing. In the pnp type, the arrow points into the flat line or base electrode. In the npn type, the direction of the arrow is reversed. Occasionally, the circle which encompasses the base, emitter, and collector leads is omitted, but this is not standard practice.

There are many other types of transistors. These are indicated in Fig. 2-45. In every case, an arrow is used in conjunction with a horizontal line and the entire work surrounded by a circle. Transistors may be made from silicon or germanium, but the schematic symbol by itself will not indicate the type of semiconductor material used.

Vacuum Tubes

While *vacuum tubes* are not used in electronic construction nearly as often as they were a decade ago, one will still

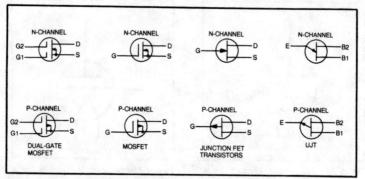

Fig. 2-45. Other types of transistors drawn schematically.

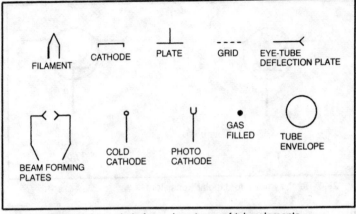

Fig. 2-46. Schematic symbols for various types of tube elements.

encounter designs which use these devices. Drawing the symbol for a vacuum tube consists of adding the symbols for the tube elements together in such a manner as to symbolize the type of tube being displayed. Figure 2-46 shows the schematic symbols for the various types of tube elements commonly used in schematic drawings. Some of these will be used in displaying every type of vacuum tube.

Figure 2-47 shows the schematic symbol for a *diode vacuum tube*. This is a two-element device containing a plate and a cathode. A filament is used to heat the cathode, so in reality, three element symbols are actually used to display this type of device. In electronics terminology, the filament is often considered to be part of the cathode. Notice that the

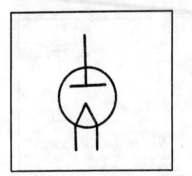

Fig. 2-47. Schematic symbol for a diode vacuum tube.

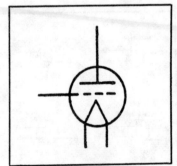

Fig. 2-48. Schematic symbol for a triode vacuum tube.

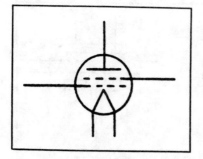

Fig. 2-49. Tetrode vacuum tube symbol.

symbols for some of the tube elements from the previous chart have been used to construct this tube symbol. All tube elements are surrounded by a circle. Occasionally, the circle will be omitted from some schematic drawings, but this is not approved practice.

Figure 2-48 shows a *triode vacuum tube,* which consists of the same elements as the diode previously discussed, with the addition of a dotted line to indicate the grid. *Tetrode vacuum tubes* have two grids, so to draw this latter device, an additional dotted line is included. This is shown in Fig. 2-49. Figure 2-50 shows other types of tube symbols that will commonly be encountered.

Some vacuum tubes actually consist of two tubes housed in a single envelope. These are known as *dual tubes* or *dual pentodes, dual tetrodes, dual triodes,* etc. Figure 2-51 shows how some of these tubes are represented schematically.

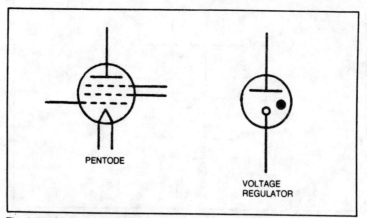

PENTODE

VOLTAGE REGULATOR

Fig. 2-50. Other schematic symbols used to indicate vacuum tubes.

33

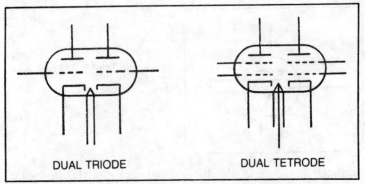

Fig. 2-51. Examples of schematic symbols used to indicate dual vacuum tubes.

A *cathode ray tube* is a special vacuum tube device and is indicated by the schematic symbol shown in Fig. 2-52. The shape of the envelope is indicated pictorially, but the internal tube elements still follow previous patterns in that plates, grids, cathodes, and filaments are shown.

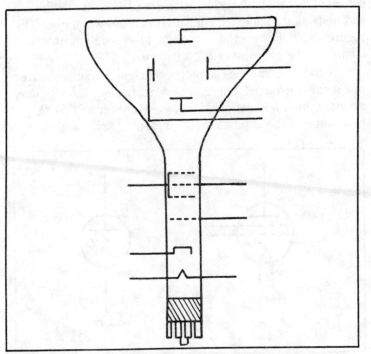

Fig. 2-52. Schematic symbol for a cathode-ray tube.

34

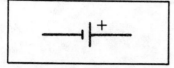

Fig. 2-53. Schematic symbol for a single-cell battery.

Fig. 2-54. Schematic symbol for a multi-cell battery.

Batteries

A *battery* is often used as a power source for many electronic circuits and must also be indicated schematically to aid the builder and serviceman. Figure 2-53 shows the schematic symbol for a *single-cell* battery. A single-cell component such as this will usually have an output of approximately 1.5 volts dc. This means that a single dry cell battery will be indicated by the drawing shown. Batteries with higher voltage outputs are usually composed of several single cells, and the schematic representation for a *multi-cell* design takes this into account, as shown in Fig. 2-54. The multi-cell symbol is simply a number of single-cell symbols combined in series. Now, if a circuit calls for the use of three single-cell batteries in a series connection, the symbol would actually be composed of three single-cell symbols in series. This is shown in Fig. 2-55. The difference between a series connection of three single-cell batteries and the symbol for a multi-cell battery can be clearly seen.

Notice that a positive and negative polarity sign are included with every battery schematic symbol. Sometimes these are omitted, so it must be remembered that the taller vertical line is always the positive connection, while the shorter one is negative. Standard practice calls for these symbols to always be included when drawing schematics. However, some artists will neglect this, and it will be up to the person reading the schematics to decipher the intent of the entire design.

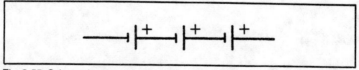

Fig. 2-55. Schematic symbol for three single-cell batteries connected in series.

"INSERT APPROPRIATE DESIGNATIONS

A-AMMETER
V-VOLTMETER
mA-MILLIAMMETER
ETC.

METERS

MOTOR

M OR MOT

BATTERIES

SINGLE CELL

MULTICELL

GROUNDS

CHASSIS

EARTH

ASSEMBLY OR MODULE

FUSE

ANTENNA

OR

MICROPHONE

HEADSET

SPEAKER

CRYSTAL QUARTZ

HAND KEY

LOGIC

AND GATE

OR GATE

INVERTER

OTHER

COMMON CONNECTIONS

CONTACTS

FEM. MALE

COAXIAL RECEPTACLE

COAXIAL PLUG

FEMALE

MALE

117 V

230 V

MULTIPLE, MOVABLE

MULTIPLE, FIXED

PHONO JACK

MIC JACK

PHONE JACK

PHONE PLUG

CONNECTORS

SHIELDED WIRE

SHIELDED MULTICONDUCTOR

COAXIAL CABLE

ENCLOSURE

GENERAL

SHIELDING

AMPLIFIER

OPERATIONAL AMP.

OTHER

LINEAR INTEGRATED CIRCUITS

NEON (ac)

PILOT

LAMPS

N-CHANNEL

G2 G1 D S

P-CHANNEL

G2 G1 D S

DUAL-GATE MOSFET

N-CHANNEL

G D S

P-CHANNEL

G D S

MOSFET

N-CHANNEL

G D S

P-CHANNEL

G D S

JUNCTION FET

N-CHANNEL

E B2 B1

P-CHANNEL

E B2 B1

UJT

pnp

C E

npn

C E

BIPOLAR

TRANSISTORS

36

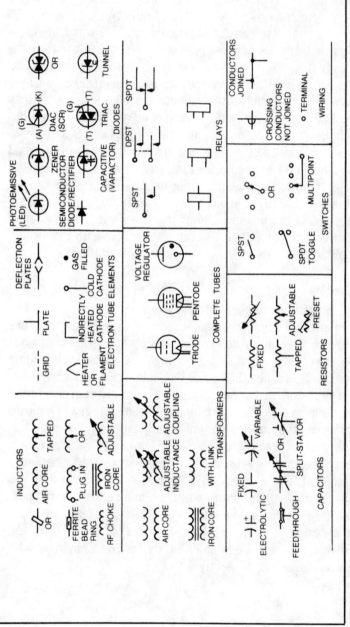

Fig. 2-56. Schematic symbols table.

37

Other Schematic Symbols

There are many other schematic symbols which are encountered in common practice. The ones discussed thus far are very common and will be used quite often. Figure 2-56 shows a complete table of schematic symbols which are commonly used in electronics applications. In addition to the ones already discussed, you will see symbols for *jacks* and *plugs, solenoids, piezoelectric crystals, lamps, microphones, meters, antennas,* and many other electronic components. It may seem like quite a chore to memorize all of these symbols, but their usage and correct identification will come with time. The best way to begin the memorization process is to read simple schematics and refer to this chart whenever a symbol crops up which you cannot identify. After an hour or so, you should be able to move on to more complex schematics, again referencing the unknown symbols. After a few weekends of practice, you should be thoroughly familiar with most electronic symbols used in schematic representations.

Summary

Schematic symbols are a system unto themselves, but most are based upon the actual physical or working structure of the component or device they represent. Schematic symbols are often presented in groupings, each of which has some relationship to the others. For example, there are many different types of transistors, but all of them are represented in similar manners. There are minor symbol changes to indicate a different type of device, but all can be easily identified as some type of transistor. The same applies to vacuum tubes. Nearly every type is represented by a circle which contains a number of tube elements. All resistors are indicated by the triangle pattern, but some will contain an extra symbol portion to represent a special usage. While schematic symbols are a system unto themselves, this is a logical system, and one that can become a quick study for the individual who pursues this field diligently.

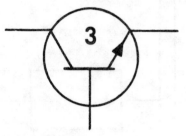

Simple
Electronic Circuits

There are two schools of thought when it comes to teaching people how to read and write schematic drawings. One school feels that the student should learn to read them before the writing process begins. The other school feels that the student should learn to read schematic drawings *by* writing them. I encompass the best of both worlds by advocating a combination of the two. Certainly, you must learn the basic symbols at the very start. After this is done, you should attempt to read as many schematic drawings as possible. When this becomes boring, you then switch to making your own drawings. If you rotate back and forth during the study period, you will probably gain a better overall knowledge. For this reason, you should devote half your study time to reading symbols and the other half to writing them.

This chapter will deal with the reading and writing of some extremely simple electronic circuits which will be shown pictorially and then schematically. Using this method, you can actually see the circuit and then see how the schematic representation is drawn from it. Some commercial schematics are produced in this manner. However, in most instances, a circuit will be designed schematically first, then built and tested from the schematic. If this is a highly experimental circuit, there will be some bugs in the test mockup

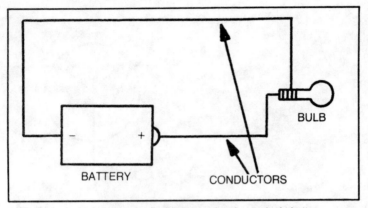

Fig. 3-1. Pictorial drawing of a flashlight circuit using a single battery.

which will require some component deletions, substitutions, or modifications. When these changes are made to the test circuit, the results are noted and the schematic is changed accordingly. In the end, the finished and corrected schematic is a product of design theory, actual testing, and modification.

Figure 3-1 shows a simple circuit that we have all used at one time or another. Basically, this is a flashlight with the external case removed. The flashlight consists of a battery and an electric bulb. This pictorial representation also shows the conductors which attach to the light bulb and the battery. The conductors form a current path between the battery and the light bulb. Current flows from the negative terminal of the battery through the bulb element and back to the positive terminal of the battery. In the pictorial representation, the positive and negative terminals are indicated.

In order to make a schematic diagram of this simple circuit, it is necessary to know three schematic symbols. These are the battery, the conductors, and the bulb, and all three are shown in Fig. 3-2. Now that we know the symbols which are needed, they can be assembled in a logical manner based upon the appearance of the circuit in the pictorial drawing.

Let's start first by drawing the battery symbol. The battery can be thought of as the heart of the circuit, since it supplies all power. Next comes the symbol for the light bulb, which can be drawn at any point near the battery. Using this

40

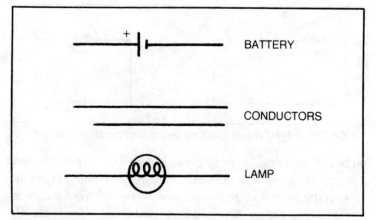

Fig. 3-2. Symbols for the battery, conductors and bulb used to make the former flashlight circuit.

example, we will try to make the schematic symbols fall in line with the way the pictorial diagram was presented. This places the light bulb to the immediate right of the positive battery terminal.

Now that the two major symbols have been drawn, it's a simple matter to use the conductor symbols to hook them together. Notice that the pictorial drawing shows two conductors. Therefore, two are used in the schematic diagram as well.

Figure 3-3 shows the completed schematic drawing which is the symbolic equivalent of the pictorial drawing we originally worked from. The schematic drawing indicated is by no means the only way this simple circuit can be represented. However, any schematic representation will re-

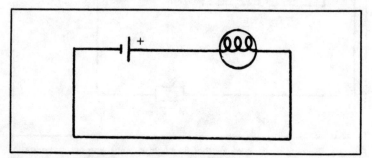

Fig. 3-8. Schematic diagram of the former circuit.

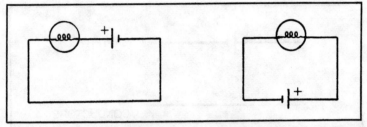

Fig. 3-4. Different methods of indicating the previous circuit schematically.

quire the use of the basic symbols outlined. The only changes that can occur involve the positioning of the component symbols on the page. Figure 3-4 shows several different methods of indicating the same circuit schematically. They are all electrically equivalent but appear different due to their relative positions. Notice that in every case the individual symbols are correctly drawn. This circuit uses a single-cell battery, so the single-cell symbol is incorporated, along with its positive and negative polarity markings.

Let's alter this circuit a bit to gain more proficiency in reading and writing schematic drawings. Figure 3-5 shows the same basic circuit, but an additional battery has been added, along with a switch. This is the most standard config-

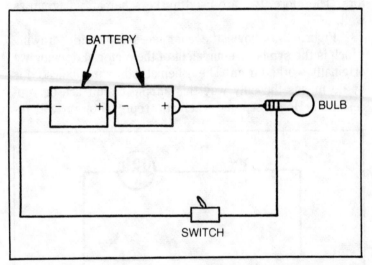

Fig. 3-5. Pictorial drawing of another flashlight circuit using two batteries in series.

uration for all flashlights sold in this country. By examining the pictorial drawing, we know immediately that any schematic representation must contain two battery symbols, the conductors, the light bulb, and the on/off switch. The switch is the only new symbol to be added to this circuit, since the single-cell battery symbol will simply be repeated. Figure 3-6 shows the symbols which will be needed to assemble an accurate schematic drawing of this circuit. Again, the symbols are drawn on paper in the same basic order as the components they represent are wired in the circuit. The resulting schematic is shown in Fig. 3-7. Note that the two single-cell battery symbols are drawn in series, with polarity markings provided for each. Using a series connection, the positive terminal of one battery is connected to the negative terminal of the other. The same two conductors are used from the battery terminals, but a third one is needed to connect the switch to the light bulb. The switch is shown in the off

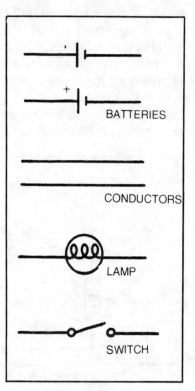

Fig. 3-6. Schematic symbols which will be used to assemble a schematic drawing of the dual battery circuit.

BATTERIES

CONDUCTORS

LAMP

SWITCH

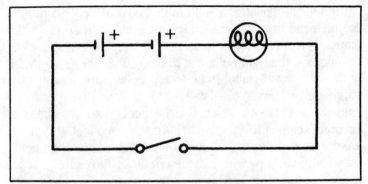

Fig. 3-7. Schematic diagram of dual battery flashlight circuit.

position, which is common for this type of representation. Congratulations! You now know what a common flashlight looks like when represented through schematic symbology.

It's time to move on to a slightly more complex circuit. Figure 3-8 shows a device which is known as a *field-strength meter*. This is a pictorial representation, and the circuit consists of an antenna, a diode, a meter, a variable resistor, and a coil. For this discussion, the values of the various components are unimportant. In order to draw this circuit schematically, it is necessary to use the symbols for antenna, diode, meter, coil, and variable resistor. These are shown in Fig. 3-9. Using the same method as before, we can draw the schematic representation of this simple device in a short period of time by

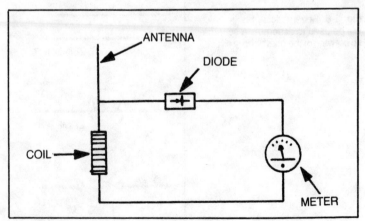

Fig. 3-8. Pictorial representation of a field-strength meter.

44

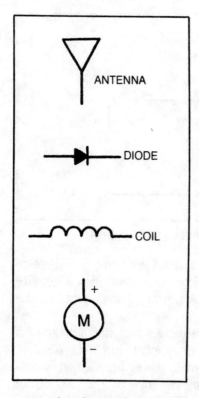

ANTENNA

DIODE

Fig. 3-9. Schematic symbols used to form schematic drawing of the field-strength meter.

COIL

M

connecting the symbols in the same order as the components they represent in the circuit.

Figure 3-10 shows how the completed schematic might look. Notice that all this involves is a simple substitution of schematic symbols for actual component parts. As before, the parts need not be physically placed in the order shown, but they must be interconnected exactly as indicated or the circuit represented will be incorrect.

By now you should be getting the idea. Previously, schematic drawings have been compared with a road map. A road map is supposed to indicate exactly what the motorist will experience in practice. The schematic drawing does exactly the same thing. We can make further analogies by saying that the highways indicated on the road map which interconnect various towns and cities are very similar to the conductors in a schematic diagram which connect electronic components. The latter are analogous to the towns and cities.

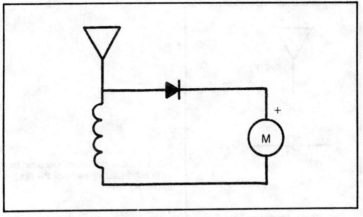

Fig. 3-10. Completed field-strength meter schematic diagram.

A road map allows the motorist to get from one point to another, whereas a schematic diagram provides a definite route for current through the various components in the circuit.

As another example of a simple schematic, a diagram is shown of an ac-derived dc power supply in Fig. 3-11. Reading from the left, we have an ac male power plug which is connected to the primary winding (the one on the left) of the transformer through a fuse. At the secondary winding of the transformer (the one on the right), a diode is connected in series. Following this, an electrolytic capacitor (note the plus (+) sign) is connected in parallel between the output of the rectifier and the bottom lead of the transformer secondary. Also connected in parallel with the capacitor and between the same two circuit points is a fixed resistor. The circuit which could be completed from this schematic could be very small or very large, depending upon the voltage and current which are to be delivered. Since dc power supplies have polarized outputs, positive and negative signs are used to indicate the output polarities. Any half-wave power supply which uses a single diode, capacitor, and resistor will look exactly like this. If the output is to be 5 volts at 1 ampere or 5,000 volts at 50 amperes, the basic schematic drawing will be identical. This is not to say that special features and additions could not be added to either supply and be reflected in the

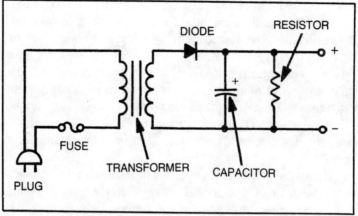

Fig. 3-11. Schematic diagram of a simple dc power supply.

schematic drawing. This states only that two dc power supplies using a very basic design will appear identical regardless of the values of the components used.

Figure 3-12 shows the same schematic drawing as before, but this time each component is given an alphabetic/numeric designation. This references each to a components list chart which is also included. Now, we can see that this particular schematic uses a transformer with a 115-volt primary and a 12-volt secondary; a diode rated at 50 peak inverse volts and a forward current of 1 ampere; a 100-microfarad, 50-volt capacitor; and a 10,000-ohm, 1-watt carbon resistor. The fuse is rated at ½ ampere. Now, we have something that is practical and can be used to build a power supply with a peak output of approximately 30 volts dc. Before each component was referenced, however, the schematic had no use in a practical sense other than to illustrate the basic components of all half-wave power supplies.

The letters used to identify each component are more or less standard. Notice that each letter is followed by the number 1. The designation T1, for instance, indicates that the component is a transformer (T) and that this is the first of this type of component referenced. This is indicated by the number 1. If two transformers were used in this circuit, one would be labeled T1 and the other T2. The numbers simply reference the proper position on the components list and serve no other purpose. The diode is referenced as D1, with D being the standardly used letter for this particular component. Standardization is not universal, however; in some instances, the diode might be labeled SR1. The SR stands for *silicon rectifier.* Some zener diodes may be labeled as ZD1, ZD2, etc. This makes little difference, however, as the component designations are written in close proximity to their symbols. If the designation D1 was replaced with SR1, there would be no doubt that the letter was intended for the symbol for a diode.

Using the example shown, it is unnecessary to include a number next to each component designation because only one of each component is used to make up the entire schematic. It is standard practice, however, to always include a letter and a number to prevent any chance of misinterpretation. In some simple schematic drawings this will not be done, but standard practice makes it mandatory. In highly complex electronic drawings, several hundred components may be used, many of which are from the same family. For instance, if you saw the designation D101, this would indicate that there are at least 101 diodes in the entire circuit and if you want to know the type and value of this particular component, it will be neces-

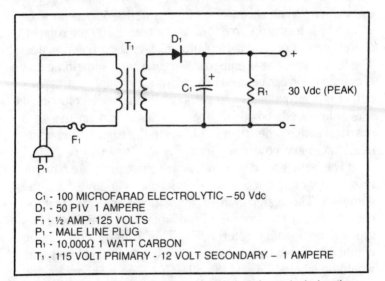

C_1 - 100 MICROFARAD ELECTROLYTIC – 50 Vdc
D_1 - 50 PIV 1 AMPERE
F_1 - ½ AMP. 125 VOLTS
P_1 - MALE LINE PLUG
R_1 - 10,000Ω 1 WATT CARBON
T_1 - 115 VOLT PRIMARY - 12 VOLT SECONDARY – 1 AMPERE

Fig. 3-12. Former schematic diagram with alphabetic/numeric designations.

ANT	Antenna
B	Battery
C	Capacitor
CB	Circuit Board
CR	Zener Diode (Occasionally, any Diode)
D	Diode
EP	Earphone
F	Fuse
I	Lamp
IC	Integrated Circuit
J	Receptacle, Jack, Terminal Strip
K	Relay
L	Inductor, Choke (Usually Audio Frequency
LED	Light-Emitting Diode (Occasionally)
M	Meter
N	Neon Lamp (Rare)
P	Plug
PC	Photocell
Q	Transistor
R	Resistor
RFC	Radio Frequency Choke
RY	Relay
S	Switch
SPK	Speaker
SR	Selenium Rectifier
T	Transformer
U	Integrated Circuit
V	Vacuum Tube
VR	Voltage Regulator (Tube-Type Usually)
X	Solar Cell (Rare)
Y	Crystal
Z	Circuit Assembly (Block Diagram Designation)
ZD	Zener Diode (Rare)

Fig. 3-13. Chart of letter designations for the various types of electronic components.

sary to reference D101 in the components list. Figure 3-13 shows the letter designations for the various types of electronic components. It should be understood, however, that these can vary slightly, depending upon the style and proficiency of the person making the drawing. Learning this chart will be quite simple, as most of the designations used are the first letters of the components they represent. If the component has a complex name, such as silicon-controlled rectifier, the first letters from each of the three name portions is used, i.e., SCR1. In this case, S1 could not be used because S is the designation for a switch. A resistor is designated by the letter R. Therefore, a component such as a relay, whose name also starts with the letter R, must have a different designation, i.e., RY1. Once you know a bit about this system of symbols and designations, the whole procedure seems much more logical.

Figure 3-14 shows a schematic drawing and components list for a more complex type of dc power supply. Many components and their basic wiring are identical to the previous power supply schematic drawing. However, additional parts have been added, and some components have been duplicated. Notice that the letter designations remain the same for each identical component, but the numbers advance in relationship to the total number of components used. Even though some components may be identical in value, they are still given seperate numerical designations, although they are combined in the components list as shown.

When building circuits from schematic diagrams, common sense must enter into the picture. It has been stated previously that single lines used in schematic drawings rep-

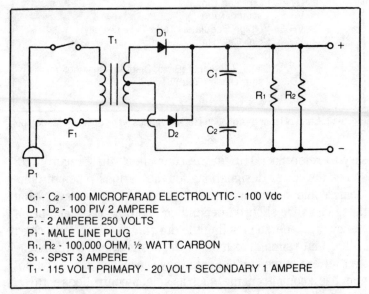

C₁ - C₂ - 100 MICROFARAD ELECTROLYTIC - 100 Vdc
D₁ - D₂ - 100 PIV 2 AMPERE
F₁ - 2 AMPERE 250 VOLTS
P₁ - MALE LINE PLUG
R₁, R₂ - 100,000 OHM, ½ WATT CARBON
S₁ - SPST 3 AMPERE
T₁ - 115 VOLT PRIMARY - 20 VOLT SECONDARY 1 AMPERE

Fig. 3-14. Schematic diagram and components list of a dc power supply.

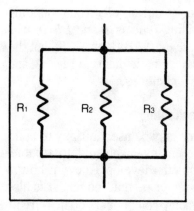

Fig. 3-15. A simple circuit made up of three parallel-connected resistors.

resent conductors. This is true in many instances. However, the conductor may not be a separate one but may actually be a part of a component lead. Whether or not a conductor is used to interconnect two components will be determined by how closely they are spaced during the construction process. I realize this sounds a bit confusing, but perhaps the schematic diagram in Fig. 3-15 will help lead this discussion to a point of clarification.

The circuit shown includes three resistors, all of which are connected together in parallel. Taking the circuit quite literally, a conductor is used to connect the bottom of R1 to the bottom of R2. Another conductor is used between the bottom of R2 and the bottom of R3. Two other conductors are used to connect the top leads of the components. In actual practice, this probably would not be done, as the three components would be mounted in close proximity to each other and the already existing leads intertwined. This is shown in Fig. 3-16. Naturally, you will want to make all electronic circuits as compact (and dependable) as possible. This involves using a minimum amount of point-to-point wiring and trying to make the component leads, which are conductors themselves, serve for interconnection purposes. Of course, in the above example, if the three resistors had to be spread out over different parts of the circuit, then interconnecting conductors would be required. This type of building expertise comes with hours of practice in building electronic circuits of a simple nature and then moving on to more complex designs.

There are many excellent books out on electronic building which are recommended reading (one is *Basic Electronics Theory—with projects & experiments*, TAB Book 1338), but the scope of this book is to teach schematic diagram reading and drawing rather than building techniques.

Using Schematics for Troubleshooting

While schematic diagrams may be used initially to build electronic devices, they are an invaluable aid to trouble-shooting equipment when problems develop. Knowing how to read schematic diagrams, however, is not enough. It is also necessary to know what electronic components do in a circuit and how various basic circuits operate. It must be remembered, though, that no matter how proficient one is at electronics troubleshooting, most repair jobs become real headaches without a good schematic representation of the equipment under test. Schematic diagrams clarify circuits. They present the various circuit elements in a highly logical and easy-to-understand manner.

When a circuit is built from a schematic drawing, it does not usually resemble the schematic physically. This was not true of the simple flashlight circuit discussed earlier, but will be true of most complex circuits. It is highly impractical to

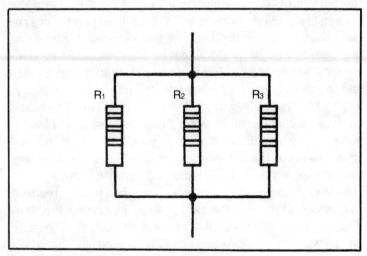

Fig. 3-16. Pictorial diagram of previous circuit.

build a complex electronic circuit by placing the components in the exact positional relationship as is done in the schematic diagram. Schematic diagrams purposely spread out the components. This is done symbolically, so the reason many schematic drawings are physically smaller than the finished device is due to the fact that the schematic symbols are physically smaller than the components they represent. Naturally, schematic diagrams are two dimensional, whereas electronic components themselves are three dimensional. One has only to look inside the television receiver or other electronic device to realize the complexities that can be involved in troubleshooting without a schematic diagram to aid you.

If you know a bit about electronic components and how they operate in various circuits, then a schematic diagram can be used to indicate (without any equipment testing) where a particular problem might occur. Then, by testing various circuit parameters at these critical points and comparing your findings with what the schematic diagram indicates should be present, a quick assessment of the problem or problems may be obtained. For example, if a schematic diagram shows a direct connection between two components in a circuit and a check with an ohmmeter reveals a very high resistance between the two, then it can be assumed that a conductor is broken or a contact has been shaken loose. By the same token, if a schematic diagram shows a 100-ohm resistor between two components and the reading with the ohmmeter is very high, this might be an indication that the 100-ohm resistor has become defective.

Beginners to electronics troubleshooting and schematic diagram reading sometimes assume that a professional troubleshooter can immediately diagnose a particular problem (i.e., locate the bad part) by simply referring to the schematic diagram. This may sometimes be true of simple circuits but is rarely so in complex designs. Often, the schematic diagram allows repairmen to make educated guesses as to where or what the trouble may be, but a true diagnosis usually requires testing. The reason for this is simple. A particular malfunction in an electronic device will not necessarily point to a single cause. Often there are many, many possible causes and then the matter is expanded from there. For example, if a circuit will not activate and no voltage can be read at any contact point as indicated by the schematic, it can be safely diagnosed that no current is getting through the circuit elements. However, what has caused this failure of the current to flow? Has one of the components in the power supply become defective? For that matter, has the line cord been accidentally pulled from the wall outlet? Is there a conductor break between the output of the power supply and the input to the electronic device? Has the fuse blown?

Here the schematic diagram may be relied upon heavily in conjunction with the various standard test procedures. The technician might wish to find the contact point which serves as the power supply output. This will be indicated on the schematic drawing. If he tests voltage at this point and it appears normal, then he can assume that the problem lies at a point further on in the circuit. The schematic diagram, along with his test instrument readings, allows him to methodically search out the problem by starting at a point in the circuit where operation is normal and proceeding forward until the point of inoperation is determined.

Using the same example, if there is no output from the power supply, the technician knows that he must search backward toward the trouble point. Chances are he will continue testing until he reaches a point of normal operation and then proceed forward from there. Using a schematic diagram, the technician may go all the way back to the original source of power, which may be the 115-Vac household supply.

All of this could be done without a schematic diagram. However, in most circuits, the length of testing would be multiplied by many, many times. As one becomes more experienced in the art of electronics troubleshooting, the information contained in schematic drawings becomes more and more valuable.

Let's return to a former circuit (the flashlight) which is shown in Fig. 3-17. While the schematic diagram does not indicate it, the two batteries in series should yield a dc

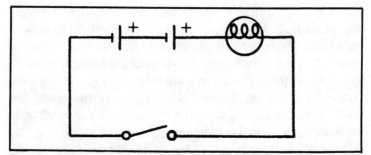

Fig. 3-17. Former flashlight circuit.

potential of 3 volts. Some schematic diagrams do provide voltage test points and maximum/minimum readings. These will be discussed later. Let's assume that the flashlight is not working and it is necessary to test the circuit with a voltohm-meter and this schematic diagram. First of all, we might measure the potential across the series connected batteries. With the positive probe of the voltmeter placed at the positive battery terminal and the negative probe at the negative terminal, we should get a reading of 1.5 volts across each battery. If both read 0, then it can be assumed that both are discharged. If one reads normal and the other reads 0, then only one will have to be replaced. However, if both batteries read normal, then the next voltage probe might be accomplished at the light bulb itself. Here, a reading of 3 volts should be expected under normal operation. If you read 3 volts here, then you should immediately know what the problem is by looking at the schematic drawing. That's right! The bulb has blown. The circuit shows that the current path is through the light bulb. If current flows through this bulb, then it has to light. If voltage is available at the base of the light bulb, then current has to flow through the element . . . unless it has opened up.

On the other hand, suppose you get a normal reading at the batteries but no reading whatsoever at the light bulb. Obviously, there must be a break in the circuit between these two circuit points. Three conductors are involved here: one between the negative terminal of the battery and one side of the bulb, another between the positive battery terminal and the switch, and still another between the switch and the other side of the bulb. Obviously, one of the conductors has broken (or a contact has been lost where the conductor attaches to the battery), or the switch is defective. Looking at the schematic, we can test for a defective switch by placing the negative voltmeter probe on the negative battery terminal and the positive probe on the input to the switch. If a reading is obtained here, then the switch is defective. If you still get no voltage reading, then one of the conductors is loose or has broken.

Admittedly, this is a very basic example of troubleshooting using a schematic diagram. But assume that the flashlight circuit is highly complex, one you know nothing about. Then the schematic diagram becomes an invaluable aid and a necessary adjunct to the standard test procedures with the voltohmmeter. This same basic test procedure will be used over and over again when testing highly complex electronic circuits. In most instances, no matter how difficult the circuit design appears to be, it is simply a combination of many simple circuits and each will be tested individually.

Figure 3-18 shows a simple electronic circuit (actually, a portion of one) which has been presented in a form that will further aid the electronics troubleshooter. The circuit consists of a single transistor and a few other components. Note that test points (abbreviated tp) have been provided at several different locations. When troubleshooting, the technician will apply his voltmeter between these points and ground and note the readings obtained. In many electronic circuits, actual voltages may deviate from design values by 15% to 20%, but this information is usually contained at the bottom of the schematic drawing. If the readings obtained are within this known error range, then the technician can tentatively assume that this part of the circuit is operational. However, if the readings obtained are 0 or well out of this range, then the technician can tentatively suspect a problem with this circuit portion or possibly other circuits which feed it. This is another method of symbolically illustrating electronic circuits and does far more for the technician than a standard diagram.

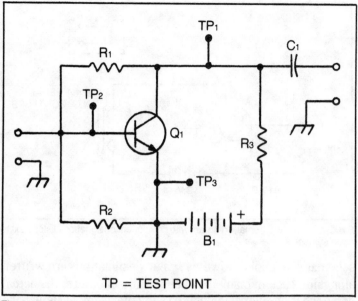

Fig. 3-18. Simple electronic circuit schematic diagram showing test point locations.

Today many schematic drawings included with electronic equipment, especially the types that are built from a kit of parts, contain invaluable information which aids not only in troubleshooting but in the original testing and alignment procedures which are often required when construction is completed. As a further aid, pictorial diagrams may also be included which serve to provide even more help in the troubleshooting process. This subject will be dealt with in a later chapter.

Other Forms of Designation

While it has been stated that it is standard practice to give every electronic component included in a schematic diagram its own alphabetic/numeric designation, there are other forms which are also pretty much standardized. Figure 3-19 shows a simple schematic diagram which does away with parts lists completely and contains no alphabetic/numeric designations. Here, the components are identified only by

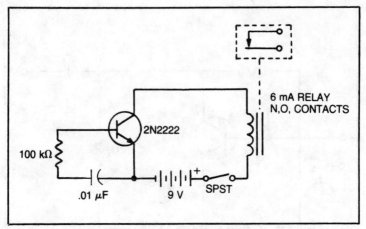

Fig. 3-19. Simple schematic diagram with component designations listed next to each symbol.

schematic symbols. However, value designations are written alongside each in many instances, while in others, the actual component number is included. Using the example shown, we know that the transistor is a 2N2222 type and that the resistor has a value of 100 K ohms (100,000 ohms). Additionally, the capacitor is valued at .01 microfarad (μF). Sometimes, a statement will be contained at the bottom of the schematic diagram which includes information about these value designations. It may read "All capacitors are rated in microfarads (μF). All resistances are given in ohms." With this type of key, the person reading the schematic diagram will be dealing with numbers only. Given the same example, the resistor would carry a designation only of 100K and the capacitor would be designated as .01. It is common practice in many schematics to use this type of designation routine. Some will even use the alphabetic/numeric system in combination with this latter system. Here a partial parts list is included which is referenced by the alphabetic/numeric designations. All components containing only a numeric value designation will be excluded from this list.

Summary

This chapter has provided a brief sampling of simple electronic circuits, showing how they can be read and/or drawn by viewing the actual completed circuit. After a bit more training, you should be able to view a simple schematic drawing and be able to visualize what the finished circuit will look like. Using the road map as an example again, we all know that a roadway does not actually appear like the black line which is used as its symbol. We can actually visualize a secondary road or even a superhighway and do so whenever we see their symbols. The same is true when reading schematic diagrams. When a capacitor symbol is indicated, we can visualize the actual component and maybe even make mental notes on the physical aspects of construction. Symbology becomes a second language, one we can begin to think in. When this level of proficiency occurs, no great mental distinctions will be made between a schematic diagram, a pictorial circuit drawing and the actual completed circuit itself.

Combining Simple
Circuits to Make Complex Ones

When beginning to learn to read and write schematic diagrams, some are often misled by the apparent complexities involved. Sure, anyone can learn to read schematic diagrams of simple circuits, those which contain a transistor or two, but it must be nearly impossible to learn to decipher more complex schematic drawings in a short period of time. This is usually not true. There is really no such thing as a highly complex circuit. Circuits which appear to be very complex are always made from building blocks, and each of these is a simple circuit. A device which contains twenty transistors and associated circuitry may really be twenty simple circuits, each using a single transistor, that have been combined. Certainly, at first glance, these scientific-looking scratchings may appear to be undecipherable, but if you just look a little while longer, you will begin to identify the simple circuit components. You will even begin to see how the simple circuits are combined.

To illustrate this point, Fig. 4-1 shows a simple circuit for a crystal radio. There is an antenna, a coil, a variable capacitor and a diode. Certainly, this could be classified as a simple schematic drawing, because the circuit is so simple that it only contains three components in addition to an antenna. If you add another component to this circuit, you can

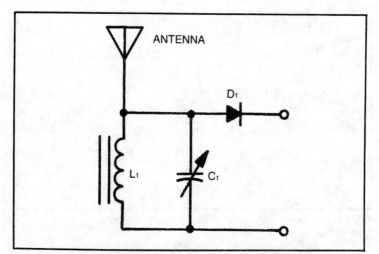

Fig. 4-1. Schematic diagram of a crystal radio circuit.

actually receive nearby AM radio broadcasts. This fourth component is a crystal headphone. However, the purpose here is not to build a simple crystal receiver, but to build an AM radio detector whose output is amplified.

Figure 4-2 shows another simple schematic diagram. This is a one-transistor audio preamplifier circuit. It consists of only six components: two capacitors, two resistors, the single transistor, and a 9-volt battery. This circuit has the ability to accept low-level audio signals as input and to pass to its output the same equivalent signal but at a higher level. This, again, is a very simple circuit, one that can be drawn in a couple of minutes and probably built in ten minutes or so.

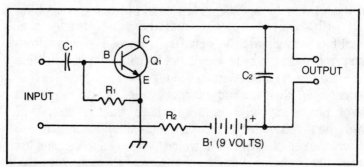

Fig. 4-2. Audio preamplifier circuit.

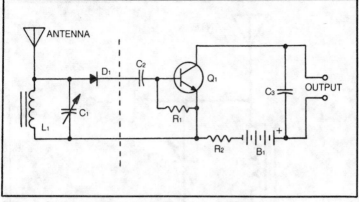

Fig. 4-3. Combination crystal radio/audio preamplifier circuit.

Figure 4-3 shows a more complex circuit. But is it really? This circuit is composed of the previous two simple circuits which have been combined to form an amplified crystal radio or a detector with a preamplifier, as it might be known. The dotted line indicates where the two circuits have been combined. If you are able to decipher the symbols in the two previous schematic drawings, you can certainly decipher this one.

The last circuit shown still has a relatively low audio output level, although it is much higher than that which would have been obtained without the transistor amplifier stage. The level would not be sufficient to allow connection of a loudspeaker, so the circuit in Fig. 4-4 can be used to increase the output signal to a point where the audio can be heard in a speaker. This is a basic example of the stages that are present in your AM table radio. This last circuit is an audio amplifier and will accept the output from the one-transistor pre-amplifier and increase the signal even more. The output from the original crystal radio would not be sufficient to drive this latter circuit, so the preamplifier stage was absolutely necessary. Notice that this audio amplifier circuit also requires a 9-volt power supply. This means that it may get its power from the 9-volt battery used for the preamplifier. Figure 4-5 shows the completed circuit, which looks even more complex but really is not, as it is a combination of three simple electronic circuits, all accomplishing different functions.

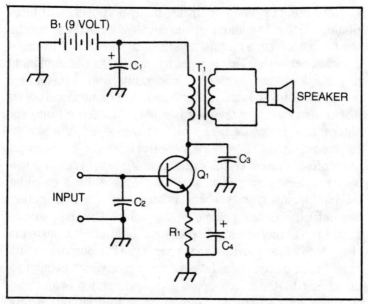

Fig. 4-4. Audio amplifier circuit.

This is the basic process used to make all electronic circuits. It's a process of combinations. First, electronic components are *combined* to form simple circuits. Then, simple circuits are *combined* to make complex circuits. Complex

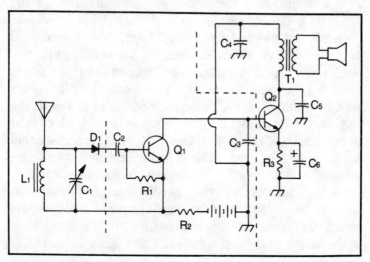

Fig. 4-5. Complete radio circuit.

circuits are combined to make multi-purpose/multi-functioning electronic equipment and devices. Even these may be *combined* to form systems. Systems are *combined* to make complex networks. This is as high as we'll take the combining process, but it can even go considerably beyond this point.

Figure 4-6 shows a rather complex-looking circuit of an AM transmitter. However, it has been blocked off into six different sections, each of which is composed of a simple electronic circuit. It may be necessary for the beginner to block off sections of circuits such as this one in order to gain a better understanding of what is taking place. Let's examine this circuit closely and see what is occurring. The first section (top left) is the microphone preamplifier. This is a simple circuit consisting of a single p-channel field-effect transistor, along with a few resistors and capacitors. The purpose of this circuit is to increase the level of the signal which is input to the circuit by the microphone. The next section is an audio amplifier which is made from an integrated circuit. A few extra capacitors and resistors are found here as well. The audio amplifier increases the amplitude of the output from the preamplifier, just as was the case in the preceding example. At the output of the audio amplifier, there is a matching network consisting of T1, a modulation transformer.

Now, moving to the bottom left of the drawing, we find the radio frequency portion of the circuit. The previous portions all dealt with audio frequency. First, there is the crystal oscillator circuit, which establishes the output radio frequency. This again is a simple one-transistor circuit which is coupled to a master network. At the output of L2 comes another amplifier stage. The crystal oscillator increases the signal established by the third overtone crystal, which is at radio frequency. This is what the microphone preamplifier did but at audio frequency. The rf amplifier boosts the signal output from the crystal oscillator, just as the audio amplifier did in the microphone circuit. Finally, there is a tuning network at the output of the rf amplifier which matches this output to the antenna. This is what the matching network did at the output of the audio amplifier, but again, at audio frequencies.

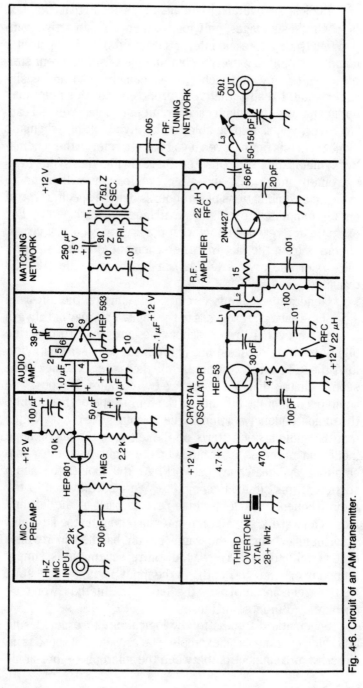

Fig. 4-6. Circuit of an AM transmitter.

65

Look at the overall circuit again. It's not all that complex, is it? Of the six stages, only four of them contain active components (transistors and integrated circuits). This apparently complex circuit is merely a combination of six different simple circuits, any of which can be quickly read and easily understood. I realize that your proficiency at this point may not enable you to explain each circuit portion as well as I can. However, you should be able to identify all of the schematic symbols in this figure, even if you have to refer to the schematic symbols table in Chapter 2. Knowing the symbols is extremely important, and after awhile, you will begin to recognize combinations of symbols as being a specific type of circuit (amplifier, oscillator, matching network, etc.). For example, the reason I know what the crystal oscillator circuit is comes from the fact that the schematic diagram shows a third overtone crystal (XTAL) at the base of the HEP 53 transistor.

Generally speaking, circuits will start on the left-hand side of the page and progress to the right. This may not always be true, especially when schematics are used to describe highly intricate devices, but is often so. This schematic follows the left or right routine. While there are six basic circuit sections in this diagram, there are two complex sections. The audio portion of the circuit (on the top) is one of them, while the radio frequency portion on the bottom is the other. Notice that the audio section starts on the left with the microphone input and progresses through to the right with the audio amplifier and matching network. At this point, the audio section of this circuit is complete. We next move on to the radio frequency section, which reverts back to the left and moves forward to the 50-ohm antenna output on the far right. You could encounter the same circuit handled in reverse order (right to left). Here, the entire schematic is simply flipped over (end over end), so the input portion would start on the right and move to the left. This method won't be encountered in most instances.

Schematic diagrams follow logical circuit order. Using this same diagram as an example, you can see that there is no direct wiring connection between the microphone pre-amp-

amplifier and the matching network. There is an audio amplifier in between. While I guess you could put the audio amplifier section in the third top block and move the matching network back to the second block, this would necessitate extending the wiring from the first block (the microphone preamplifier) to the third block, temporarily bypassing the second block, which would then be connected to the output of the audio amplifier in the third block. This makes for highly complex drawings and diminishes intelligibility. A schematic diagram is supposed to clear up misunderstandings, not add to them. The exact same comparison can be made to the rf section below.

The schematic diagram usually follows the logical amplification or processing order of the input signal. This simple transmitter receiver its audio frequency (voice) input from a microphone. This signal is immediately amplified by the microphone preamplifier circuit, so it fills the first block. Next, the preamplifier-boosted signal must be further increased by the audio amplifier circuit, so this one is next in line. After final amplification, the output impedance of the audio amplifier must be matched, so the matching network fills the third block. You see, it's all logical if you take the time to understand the schematic drawing process and how it relates to the circuit it so efficiently describes.

Figure 4-7 shows a schematic diagram which indicates only a small portion of the circuitry that makes up Hewlett-Packard's Model 7802B Monitor, a piece of biomedical equipment used to monitor heart conditions. This is one of those devices which is made up of a large number of complex circuits in order to form a system. The total schematic diagram of the entire circuit would be contained on many, many pages. The circuit portion shown is composed mostly of integrated circuits, but there are also a few discrete transistors, along with a good sampling of diodes, resistors, capacitors, and other electronic devices. Notice, however, that each basic circuit section is broken down as in the previous example. The schematic is logically arranged from left to right. The inputs are on the left, while the outputs are on the far right. The different circuits are broken down and labeled

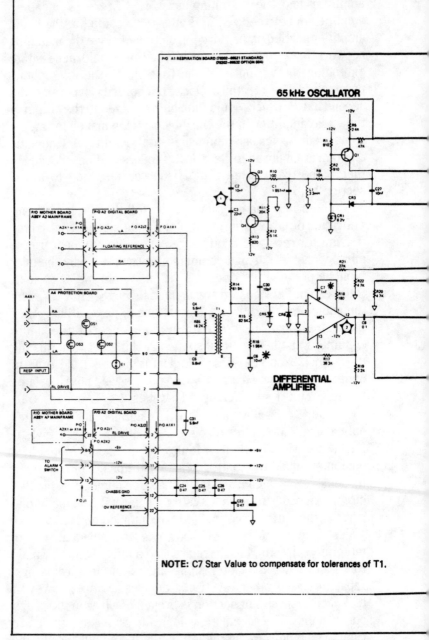

Fig. 4-7. Schematic diagram of a portion of the Hewlett Packard arrhythmia monitor (courtesy Hewlett Packard Medical Division).

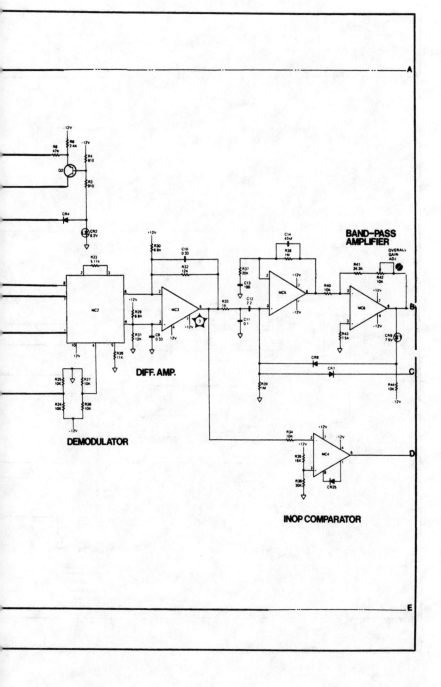

BAND-PASS AMPLIFIER

DIFF. AMP.

DEMODULATOR

INOP COMPARATOR

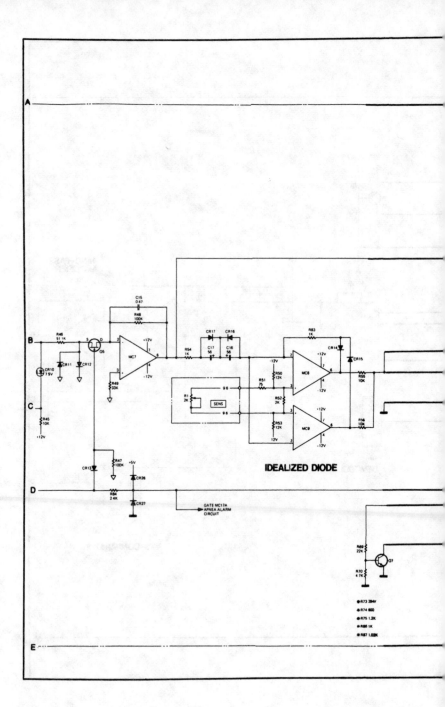

IDEALIZED DIODE

GATE MC17A
APNEA ALARM
CIRCUIT

⊕ R73 294V
⊕ R74 600
⊕ R75 1.2K
⊕ R86 1K
⊕ R87 1.02K

70

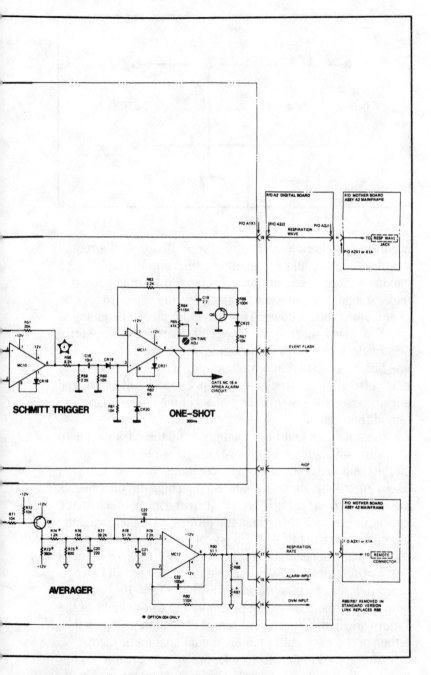

Fig. 4-7. Continued from page 70.

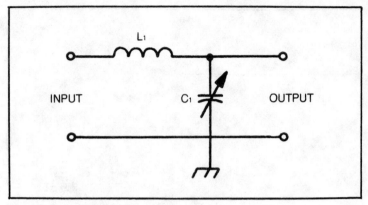

Fig. 4-8. An L-network.

accordingly. There is a 65-kilohertz oscillator, a differential amplifier, demodulator, another differential amplifier, a bandpass filter, a comparator, etc. This allows for much better understanding and an easier service routine, should a section of equipment break down. This particular piece of equipment contains many, many printed circuit boards and the entire schematic shown lists components which are found on only one board. The solid line broken by two dotted ones around most of this circuit indicates the printed circuit board and encompasses the schematic symbols of the components contained thereon.

Lest it seem that I have jumped from the ultra-simple to the ultra-sophisticated, let's return to some other simple circuits and see how they can be combined to make complex ones. Figure 4-8 shows an antenna matching circuit which is known as an *L-network*. It derives its name from its schematic appearance, which resembles the letter L. Another matching circuit is shown in Fig. 4-9. This is called a *pi-network* because its schematic appearance resembles the Greek symbol for pi (π). Both of these are very simple circuits, but can be combined to form a *pi/L-network* as shown in Fig. 4-10. Here, the output from the pi network is simply fed to the input of the L-network. This latter circuit seems to be more complex than either of the two, but is simply a combination of them.

Figure 4-11 shows a code practice oscillator circuit which is designed to be operated from a 9-volt battery. Notice

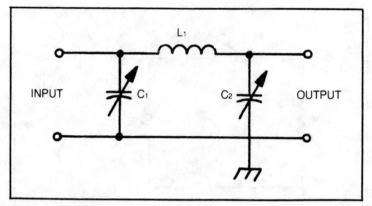

Fig. 4-9. A pi-network.

that the key symbol is used to indicate the code key, there is a single transistor whose circuit is composed of a few additional components, and then the 9-volt battery. However, suppose it is desirabie to replace the 9-volt battery with a 9-volt dc power supply which can be built from the circuit shown in Fig. 4-12. Here, we have two discrete circuits. One is an oscillator and the other is a power supply. Both can function independently, although the oscillator will require a 9-volt battery in this case. However, Fig. 4-13 shows how the two have been combined to apparently form one complex circuit. This same schematic could also be presented as shown in Fig. 4-14. Here the two circuits are included on the same page but are kept

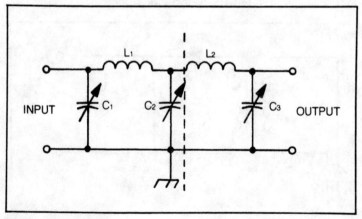

Fig. 4-10. Pi/L-network.

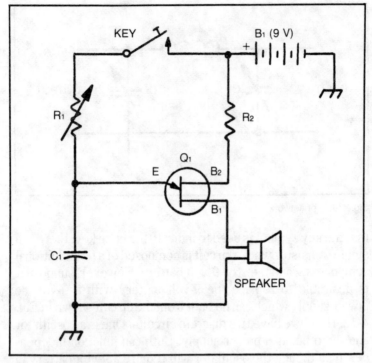

Fig. 4-11. Code practice oscillator circuit.

separate by spacing. A cable with matching connectors is used to allow the power supply to provide current to the oscillator. This latter circuit may be a little less confusing, but it is electrically equivalent to the previous one.

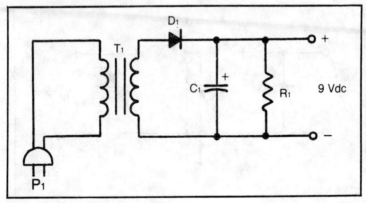

Fig. 4-12. Nine-volt dc power supply.

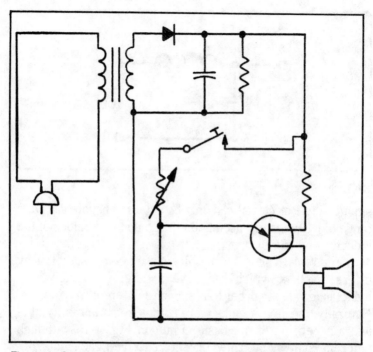

Fig. 4-13. Combination of the code practice oscillator and dc power supply circuit.

Figure 4-15 shows a filter circuit which has an input and an output. The inductor and capacitor (in practice) will be chosen to present a certain value that would allow some signals to pass and others to be blocked. Since this is only a

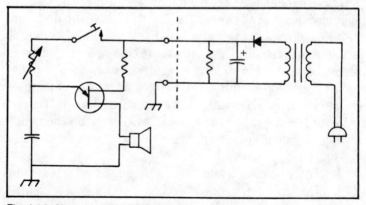

Fig. 4-14. Alternate method of representing the previous circuit.

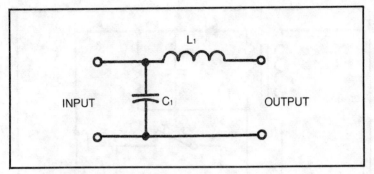

Fig. 4-15. A simple filter circuit.

simple single circuit, it can only be set up to handle a certain range of frequencies based upon the component values of the inductor and capacitor.

Figure 4-16 shows a complex circuit composed of many filters, each of which can have a different capacitance/ inductance value in order to respond to different frequencies. Naturally, the circuit looks more complex, but in reality, it is just a repetition of the basic circuit in Fig. 4-15, although different component values may be chosen for each inductor/ capacitor combination. You will find this in many electronic circuits and their equivalent schematics.

Sometimes it is necessary to have three or four of the same circuits arranged in parallel or series to bring about a particular electronic result.

If you know how one operates, then you know the basic operation of them all. A problem that can occur in one circuit might also occur in all the others, and the schematic diagram can be used for tracing purposes. For example, if through testing you know that a particular oscillator changed frequency because of a defective grid resistor, then should another oscillator exhibit the same characteristics of improper operation, the schematic can be consulted, the grid resistor located, and a further test made. Without the schematic drawing, it would be very difficult to find the grid resistor quickly.

It can be further stated that this single grid resistor which might cost ten cents or so could be causing a highly complex system to fail. By isolating a complex system of circuits into a number of complex circuits and then breaking down the complex circuits into simple circuits, schematically-aided troubleshooting becomes less difficult.

Breaking a system down into complex circuits, makes it easier to determine which circuit may be creating the problem for the entire system. This complex circuit is then broken down into simple circuits, and a determination is made as to which simple circuit might be at fault. The simple circuit is then broken down into separate components, which are then examined individually for a possible fault. Through this systematic method of elimination (all done schematically), the equipment is repaired.

I have repaired some highly complex pieces of electronic equipment (everything from biomedical instruments to television transmitters) and have found that, by and large, most failures result in a problem with a single component. Sometimes this will cause other components to fail as well, but the repair must start with the first failure. Occasionally, two simultaneously defective components will be discovered, but

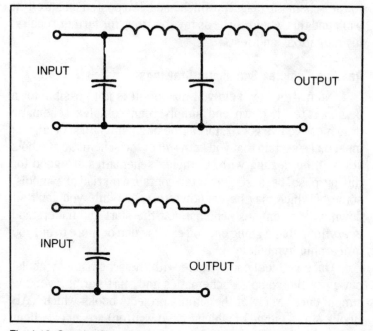

Fig. 4-16. Complex filter circuit.

this is a rarity. Certainly, it is necessary to become familiar with equipment you are attempting to repair, but after normal operation is studied and understood, a schematic drawing can often be used to identify the general area of any future faults. From this, highly suspect discrete components may be identified. It is then a matter of getting into the equipment with a meter or other test instrument and checking these suspicious components. Even with the finest test instruments, it is almost impossible to quickly identify defective sections and especially components without a schematic diagram, because you simply don't know where to look. As a long-time commercial radio/television engineer, I quickly learned the value of schematic diagrams. As a matter of fact, I often had the schematic drawings of transmitters and other indispensable equipment blown up to a very large size so that they could be hung on a wall near the equipment to serve as a constant reference source. Some transmitters are very large, and it becomes necessary to actually walk inside to perform repairs. When a doubt surfaced, I could simply look over my shoulder to the schematic diagram which had been posted on a well-lit wall, and this would serve as my road map for further repairs, investigations and tests.

Reading Complex Schematic Drawings

No matter how avid you may be, it is not possible for a beginner to sit down and simply read complex schematic drawings. You've got to start at the beginning. First of all, you must make certain that you know *all* of the schematic symbols you will be dealing with. Complex schematics are good for this purpose, because they often contain a myriad of symbols, some of which may be unknown to you. While you can't sit down with a complex schematic at the start and understand everything that's going on, you can use one or more to aid you in learning symbology.

Once you feel comfortable with the schematic symbols, put away the complex schematics and start poring over the simple ones. Many of the simple projects books which TAB Books offers (some of which I have written) are an excellent source. However, you will also find them in magazines geared to the electronic enthusiast, especially the beginner. Don't study just one type of schematic (those which deal with oscillators, for example). Peruse many. Check into oscillators, amplifiers, solid-state switches, tube-type circuits, rf circuits, audio frequency circuits, etc. You will find that many of them are basically alike, with only a few changes in component values. When you can identify an amplifier circuit by looking at the schematic alone, then you are making progress. It is impossible to learn to read and to draw schematic diagrams without learning a great deal of practical and theoretical electronic design and application. Too many people feel that all one has to do to become a crack theorist or troubleshooting technician is to learn to read schematic drawings. This is not true. Just because you know how to read a road map doesn't qualify you to drive an automobile from coast to coast. It is necessary also to learn how to drive a car. The schematic diagram is an aid to the understanding of electronics and electronic circuits. Like the road map, throughout your electronic pursuits, the schematic diagram will be a constant aid, giving you an indication of what could be wrong and where it is found in the circuit.

Once you can comfortably identify simple electronic circuits from schematic diagrams, it is time to move on to more complex drawings—not *too* complex, however. If you try to move too quickly, you may become frustrated and give up altogether. Your next step will be an intermediate one and will include the circuits which combine a *few* of the simple circuits you have previously studied. Sometimes additional components are added in order to match the output of one circuit to the input of another.

This involves a new phase of your course in reading schematic diagrams. Try to select books and publications which offer a simplistic, theoretical, and practical discussion of the circuit which the schematic depicts. Again, electronic projects books are ideal, because they usually include a schematic diagram along with an explanation of the basic circuit functions.

Even better, begin to build some simple circuits in a workshop at home. Many books have been written about building electronic circuits, but most projects books include the basics in the front and the circuits in the back portion, so these are quite handy. You will be surprised at how your first electronic circuit looks as a physical reality when compared with the schematic drawing. Your study continues from this point by examining the functioning circuit and noting the relationship of the physical components to those in the schematic drawing. You can further your electronic knowledge by experimenting with these circuits, substituting different components, for example. You may even be able to bring about an improvement in circuit operation. All improvements could be duly noted and a new schematic can be drawn up indicating the changes you effected. You may want to simply pencil in the changes on the schematic diagram you were building from.

Now, when you feel comfortable in building electronic circuits from simple schematic diagrams, you may want to combine two or more into a single circuit. Take two schematic diagrams from a projects book and combine them on paper. This means that you will have to draw your own schematic which will serve as your plans for the building procedure to follow. You may even be knowledgeable enough this time to provide some additional circuits which may be needed to interface the two. Combining electronics building with learning to read schematic diagrams is the most efficient way to improve your electronics knowledge. Admittedly, it can be quite boring to pore over schematic diagrams for hours on end. However, when you are in a position to refer to a portion of a schematic drawing and then wire the components in place, much of the boring aspect is removed and knowledge is more efficiently retained.

Before you know it, you will have obtained a great deal of basic knowledge regarding the schematic diagrams *and* the building of electronic circuits. The circuits which you used to think of as being complex now seem like old friends. *Caution:* At this point in your development, you may have a tendency to stick only with the circuits you know best. Don't let this happen. As soon as you reach one stage of comfort, move on to the diagrams which are more difficult and make you feel uncomfortable again. If you don't do this, you will be stuck at this one stage of development and stuck there for a long time. While it may not be necessary, try to continue building the more complex projects. Admittedly, this can become expensive, so if building is not possible, continue to read them and decipher the various circuit components anyway. Leave no leaf unturned. You will continually be surprised at what you know, and what you don't know. For instance, many people fairly new to electronics feel that a commercial AM radio transmitter must be a highly complex device. Most are surprised to learn that it is technically less complex than the transistor pocket receiver you use to detect the broadcasts. By comparison, a commercial transmitter is a simple circuit even though it may be seven feet high and just as wide. This is due to the fact that a modulation transformer, for example, in a commercial transmitter may weigh several hundred pounds. This and other components make this equipment necessarily large. However, a modulation transformer for a walkie-talkie may weigh less than an ounce. Schematically, both modulation transformers will be drawn in an identical manner. Circuit complexity, then, has nothing to do with the physical size of a component. It has to do with the number of components as well as the number of different combinatorial circuits included. A schematic diagram presented earlier in this chapter showed the circuit of a simple radio transmitter. The circuit of a commercial radio transmitter would not look too much different, although the components specified might be a hundred times the size of the ones used for the previous circuit.

No, it's not usually the massive pieces of equipment which are the most complex, both electronically and schematically. It's those tiny units which can be held in the palm of the hand that often take the prize for schematic complexity. Due to this, the beginner to electronics and schematics should not shy away from any particular circuit, device, or equipment just because there is an impression there of tremendous complexity. You may be wrong, but even if you aren't, there will be portions of every schematic diagram that you will be able to comprehend. After you have passed the intermediate stage of learning schematics, then the complex circuits are next in line. The first thing to do is to break them down into intermediate circuit sections and then again into simple circuits. Try to obtain schematics of a complex nature which also include a thorough explanation of how the circuit functions.

Summary

Reading and drawing schematic diagrams involves the methodology of breaking down complex circuits into simple ones. This is the way even the seasoned professional goes about this procedure. The beginner must do the same thing, looking at the part rather than the whole. As a complex schematic drawing is systematically studied, the relationship of one circuit to all the others will eventually become apparent.

When using schematics for electronics troubleshooting, it is often unnecessary to understand the function of all circuits, only the ones which are potential trouble spots. Learning to read and write schematic diagrams is very similar to learning to receive and send Morse Code. Morse Code is a language of audible symbols, just like schematic drawing is a language of printed symbols. Once you learn either language, you are capable of communicating in it and allowing it to communicate with you. Using Morse Code as a further example, a five-minute sequence of dots and dashes will mean nothing unless you are capable (initially) of breaking the data down into individual code words. As proficiency increases, the person deciphering this code stops hearing mere dots and dashes and begins hearing letters instead. It's at this point that you begin to think in the new language. As proficiency continues to increase, entire words are heard instead of the letters which form those words. Eventually, one will hear entire sentences and finally, it's just like a second language.

Reading and writing schematic diagrams follows a similar pattern of development. At first you will see individual components. Later, you will begin to see complete simple circuits hidden within complex circuits. Then, complex circuits can be identified and deciphered at a glance. Finally, you will begin to envision entire systems. This will not come quickly, but your proficiency will improve *every* time you practice if you push yourself into areas which are continually unfamiliar to you. This is a step-by-step process, one that proceeds as fast as the student is able to absorb the information presented.

Schematic Drawing Techniques

Most of the discussion to this point has dealt with reading schematic diagrams. Of course, you have been continuously encouraged to draw them whenever possible. This chapter will deal exclusively with how to draw appropriate schematic diagrams in a form which other electronics enthusiasts can read and appreciate.

The purpose of schematic diagrams is to allow electronic circuits to be represented on paper by means of symbology. Why symbols? Because they are easier to draw than pictorial diagrams and require far less artistic talent. Anyone who can write numbers can also draw schematic diagrams. Most symbols are composed of straight lines and/or circles and can be compared with the stick figures we all drew in grade school.

Up to this point the schematic diagrams in this book were all drawn by professional artists who specialize in this type of work. Realistically speaking, you will not be able to produce schematic diagrams as exacting as those produced by professional artists. This should be of no concern, however, because hand-drawn schematics (if done properly) will be as easily readable, although they will lack some of the finer artistic points. I am speaking here of schematic diagrams which are drawn freehand. If you find it necessary to produce "perfect" schematic diagrams, a discussion later on in this chapter will give you some guidelines on the proper approach to take.

The Drawing Table

When you start the business of drawing schematic diagrams, it is essential that you provide yourself with a comfortable workspace which is well-lit and provides good freedom of movement. The top of a table will do just fine, although if you have access to a slanted drafting table, this will be even better. Your setup should include a comfortable chair which will keep you from becoming fatigued during long periods of practice. The lighting is most important and should provide good illumination of the drawing surface without creating glare. Eye strain (along with writer's cramp) can be a real problem when several hours are to be put into a schematic drawing assignment.

Figure 5-1 shows the homemade drawing table I use for schematic diagram work and general drafting. It was built with an intentional slant, which makes the whole process a bit easier. If you don't have one of these, don't go to the expense of purchasing or building one unless you intend to do a great deal of schematic diagramming. The light over my table is a cool-white fluorescent type and provides even illumination without glare. Chances are, you will need a very bright incandescent light if this type of bulb is to be used. This presents a problem with glare, so the fluorescent type is to be preferred.

Naturally, you will want to have a good supply of sharpened No. 2 pencils and heaping quantities of paper. I use inexpensive typing paper, but many other types will be suitable. This is often called *copy paper* and will be sold at your local stationary store. Do not choose a paper with a high-gloss finish or good bond paper. The latter is quite expensive and its textured finish is not the most desirable for schematic diagramming. You should also have on hand a plastic ruler, and possibly, an inexpensive compass for making uniform circles.

This is about all you'll need to start out. It should be understood that schematic diagramming is often done on the back of a paper bag with a nub of a pencil and under the most adverse conditions. On occasion, I have produced a readable

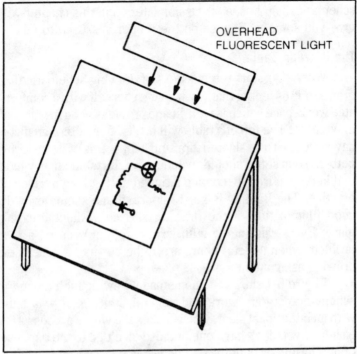

Fig. 5-1. Homemade drawing table.

schematic diagram in some loose dirt with a stick during field conditions when a repair operation was taking place. As you become more comfortable in drawing schematic diagrams, the facility previously described will not be as necessary. However, this applies to drawing diagrams for yourself. When schematics must be provided for other persons, you will still want to return to your professional setup to do the best job possible.

The schematic diagrams presented in this chapter are exact reproductions of my drawings, done in a more or less freehand style. This will give you a better idea of the results you can expect to achieve in a practical manner at home. The quality of these drawings will not be as good as those which can be produced using specialized schematic drafting equipment, but all components and their interconnections can still be clearly defined to produce a pleasing end product.

There are a few additional accessories that you might

wish to have on hand. Make certain that you have a good eraser or two. This item should be cleaned after each use by rubbing it across the surface of a sheet of white paper. Erasers remove pencil markings by lifting the graphite off the page. This material then remains on the surface of the eraser and may be transferred back to an area of erasure next time the eraser is used. By using the clean sheet of paper (one that contains no graphite markings), you can remove most of the graphite contents of the eraser by simply rubbing it a few times across the paper's surface.

A few coins can also come in handy for making circles. These will later be filled in with the correct line symbols to produce transistors, silicon-controlled rectifiers, etc. It's a simple matter to place the coin on the paper and then trace around its edges. Of course, if you have a compass, this may not be necessary.

While schematic drawing done with a fine, black ink pen is more legible than one drawn with pencil, this should be a pursuit that is saved for much later. All schematic diagrams should initially be done in pencil. Once all of the problems of schematically representing a specific circuit are worked out in this medium, you may choose to redo your finished design in ink. If you intend to do this, a bottle of correction fluid will be helpful.

Finally, you will want to post a schematic symbols chart directly in front of you (preferably on a wall) so that you can make constant reference to it. This may not be necessary after a while, but for right now it will come in very handy. You can use this chart to supply you with the symbols you cannot remember. It should also be used to check the accuracy of symbols which have already been drawn.

Drawing Symbols

The time has come to put some of the knowledge you have attained into actual practice. Begin by drawing every schematic symbol on the chart in Fig. 2-56. Try to make all of your symbols of comparable relative size. For example, the symbol for a transistor is usually made larger than that of a resistor. Figure 2-56 displays all component symbols in proper relationship to size. Figure 5-2 shows a few of the symbols as I actually drew them. Notice that there is no question as to what the symbols are and the components they represent, but they do not look *exactly* like those in Fig. 2-56 drawn by a professional draftsman using specialized equipment.

Too many beginners to schematic diagramming neglect practicing the art of drawing the individual component's symbols. If you can't draw these correctly, then your schematic diagrams will not be correct and will consist of a hodgepodge of component symbols which may be vague or mean little or nothing. I realize that you may be anxious to get on with the actual diagramming of entire circuits, but these basics cannot be ignored. Your first day of practice should be devoted to nothing but drawing schematic symbols.

Let's take the symbol for a pnp transistor as a good example of how to begin forming the various components. Figure 5-3 shows the step-by-step process which includes:

1. Forming the circle with a coin or compass.

2. Drawing the horizontal base line inside and just below center of the circle.

3. Drawing the vertical base lead.

4. Drawing a diagonal line to the base line for the collector lead.

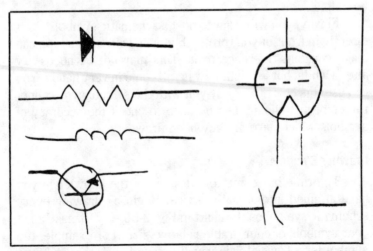

Fig. 5-2. An example of schematic symbols as they appear when drawn freehand.

88

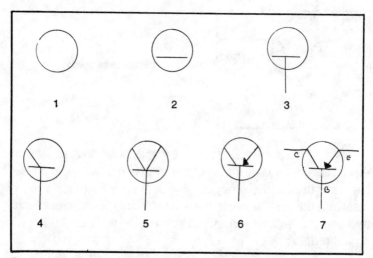

Fig. 5-3. Step-by-step method of drawing a transistor symbol.

5. Drawing a mirror image diagonal line on the opposite side of the transistor base line.

6. Drawing and filling in a small arrow at the end of this last diagonal line to complete the emitter lead.

7. Drawing the identification letters next to each electrode line (optional).

While this process may initially take a bit of time, after a few minutes of practice, you will become accustomed to how the pnp transistor symbol is formed and will be able to do it in a much shorter period of time. The coin or compass is used to draw the circle and the straightedge comes in handy for forming the straight lines which indicate the various electrode leads. The arrow will have to be drawn freehand in most instances. If you wish to draw an npn transistor symbol, as shown in Fig. 5-4, the above process would simply be repeated, with the exception that the arrow would be pointing in the other direction and located at the top of the emitter line as indicated.

A resistor would be drawn following the steps shown in Fig. 5-5. The three triangles are put down on paper freehand, and then the straightedge is laid along the longitudinal center of this portion of the symbol. The horizontal end leads of this symbol are drawn using a straightedge.

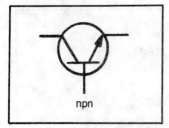

Fig. 5-4. Npn transistor symbol.

Vacuum tube symbols are, for the most part, drawn in much the same manner as the pnp transistor. Once the circle has been drawn, the straight edge may be used to aid you in inserting the various electrodes. Even the broken lines which indicate grids can be drawn more easily with the aid of a straightedge.

Undoubtedly, drawing many of these simple symbols will be easy for you, but there will be some which will be a bit more difficult. The latter are the ones you should especially concentrate on. Draw several of them and then compare them with what is shown on the symbols chart. How do yours differ from those that are contained here? Keep practicing until your symbols closely match those presented on the chart.

Pace yourself during all of these exercises, because it's quite easy for fatigue and the frustration associated with this and other exacting jobs to set in. When you first start, draw symbols only for about fifteen minutes at a time and then take a break for five or ten minutes. This is especially true if you're not regularly employed in a job which requires the handling of a lot of paperwork. You don't need to quit your studies completely, as you can still study and read other schematic drawings while you're resting your hand. The total length of time which you can devote to practicing schematic diagramming will depend upon you and you alone. I am speaking here of

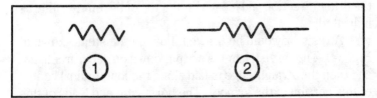

Fig. 5-5. Steps used to draw a fixed resistor.

your efficient mental attention span now and not the time you may have available. You may find that your first few days of practice are quite difficult, but after a while you will become more proficient at what you're doing and will be able to spend more and more time with this practice.

Drawing Simple Circuits

After you feel comfortable drawing the many different types of schematic symbols and can do so repeatedly without having to refer often to the symbols chart, then the time has come to move on to some simple circuits. At first, choose a circuit which has already been schematically diagrammed in a book or magazine. Place it in front of you and then attempt to reproduce it without tracing. This will give you practice in knowing how to properly space the various component symbols from one another and interconnect them. Figure 5-6 shows a professionally done diagram of a simple circuit, along with the one the author made from it on his own. Notice that there are slight differences in symbol size, spacing, etc., but there can be no doubt that these two diagrams portray the same circuit.

To many students, it would seem that if you can draw individual symbols properly, combining them to make simple circuits should be no problem at all. This is a mistaken assumption. Sure, all you need to connect the components is a straight line or two, but look at the schematic again. Are the lines really straight? No! They are composed of straight

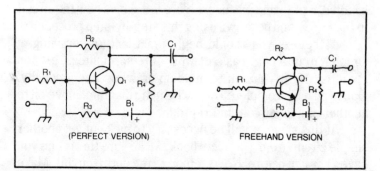

Fig. 5-6. An example of a professionally drawn electronic circuit and its freehand equivalent.

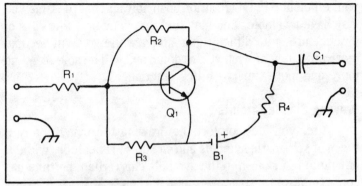

Fig. 5-7. Incorrect method of drawing previous circuit.

sections, but there are several 90° bends which allow a single conductor symbol to be routed to various portions of the schematic drawing. Figure 5-7 shows this same circuit drawn in a manner which is quite common to beginning students. The lines do not bend at right angles and twist and turn every which way. The diagram still portrays the same circuit, but it is far more difficult to read.

Usually, a simple or complex schematic diagram will begin by drawing the symbols first and *then* interconnecting them with associated wiring. Highly complex schematics may require that this be done in circuit sections. For example, the schematic drawing of a radio transmitter might start out by including only the symbols for the components in the oscillator section. These would be connected by wiring symbols before moving on to the next circuit section, which would be handled in the same manner. Figure 5-8 shows how the circuit in Fig. 5-6 would be drawn using this step-by-step procedure.

As before, you should begin practicing the drawing of simple circuits on a basis of work fifteen minutes, rest ten minutes. Allow yourself to build up drawing time gradually. Once you have completed one schematic diagram, move on to another simple circuit and reproduce it.

At this stage, it will be necessary to pick out yet another simple circuit from a projects book. Make sure this is one you haven't drawn before. Now memorize the entire circuit. Make sure you choose a diagram which will not require hours and hours to memorize. A good start might be a simple oscillator circuit that uses a single transistor and a few other components. Once memorized, you can now begin to draw it on paper. Since you won't have a visual copy to work from, this will be a little more difficult and will require you to space components properly based upon your knowledge of the relative space each will occupy. This is one of the most asked about problems in schematic drawing. "How do I know how much space to allot for each component?" This is something that comes with practice and can be described as a "feel" for schematic diagramming. This feel will come in mighty handy at a later time when you may find it necessary to draw a complex circuit based upon memorizing several simple circuits. Here, spacing is critical to the completion of a correct and easy-to-read diagram.

Repeat the procedure of memorizing a simple circuit and then redrawing it from memory over and over again until you begin to develop this feel for schematic diagramming. This will probably be the most difficult phase of your training, but if you stick to it, it will be the most rewarding as well. Should you encounter severe difficulties, then there's a good chance that you have not worked long enough in practicing the two previous steps to reach this level. Go back over these methods again until you are certain that you are ready to move on.

When learning schematic drawing, a certain amount of time should be set aside out of each day for practice. You will progress at a much slower pace if you only practice a day or

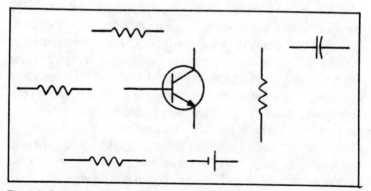

Fig. 5-8. Step-by-step method of drawing previous circuit.

two out of the week. As has already been stated, memory of symbols is involved, along with the development of a feel for schematic drawing. A reasonable level of proficiency can only be attained through daily practice routines. If you skip a week, then you will most likely lose some of the proficiency you have built up previously.

Towards More Professionalism

To this point, the discussion on drawing schematic diagrams has dealt with the methods most persons use in diagramming for themselves and others on a non-professional basis. This is a more or less freehand style, although a few implements are brought into play. After you become accustomed and comfortable with schematic diagramming, you may even do away with the compass and straightedge and still produce fairly neat drawings.

If you wish to be able to produce highly professional diagrams, you will need many more hours of practice along with some specialized electronic drafting tools. Figure 5-9 shows an inexpensive accessory which may be used by the electronics draftsman. It is a combination straightedge and electronic symbols template. These are often made of plastic or light aluminum and cutouts for the various symbols are made at various points along the device. There are circles of various sizes to form the symbols for transistors, silicon-controlled rectifiers, and other components which require their electrode element symbols to be housed within a circle. The electrode elements for pnp, npn, and even MOSFET transistors are contained in separate cutouts. The trick here is to draw the circle first and then choose the appropriate size of electrode cutout to fit the center. Additionally, symbols for antennas, resistors, capacitors, and many other electronic components are included. The tool shown in this figure is very simple and quite inexpensive. More professional outfits may be purchased which include separate templates for every electronic symbol. These can provide multi-sized symbols. For example, there may be three pnp transistor templates, each of a different size. The size chosen will depend upon how large or small you wish your total schematic diagram to be.

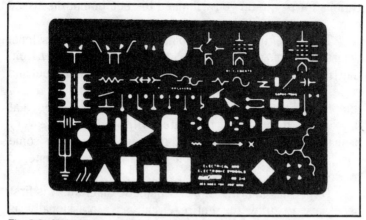

Fig. 5-9. An accessory tool used to make schematic drawings.

The purpose of templates and other implements used to make professional electronic schematic diagrams is to assure uniformity. Since all symbols will be placed on paper by means of a template, all of one type will look exactly the same.

We have all used templates at some period in our lives. Chances are in grade shcool, you used alphabetic templates to draw words on posters. These are simply cutouts in the form of letters, numbers, symbols, or whatever that are laid on the paper and traced. Schematic diagramming templates, however, can be quite expensive, costing $250 or more. Again, these are only necessary when it is essential that professional schematic diagrams be produced. Schematic diagrams which are drawn using templates and other specialized tools take considerably longer to finish than those which are done freehand.

If it's necessary to sketch a schematic from memory in order to visually see the symbolic circuit for troubleshooting purposes, you probably would not want to go with the extra time and expense that would be incurred in using the template method. Likewise, if you have a favorite electronic circuit that a friend desires the schematic for, he certainly wouldn't need one that was in textbook form. You would simply sit down and sketch a neat rendition freehand. However, if you hope to advance in this pursuit to a position whereby you can produce schematic diagrams for local electronics companies or publi-

95

cations for pay, then you will need the equipment discussed here.

To many, the use of templates to produce professional diagrams seems quite simple. Nothing could be further from the truth. This method is more difficult than the freehand style. All interconnecting lines must match up and the use of templates is a practiced art. Chances are, your first attempt at drawing with the aid of templates will produce some very haphazard results. Like any tool, the template requires some special procedures and using it properly can't be mastered overnight.

To produce a professional schematic diagram, it is first necessary to draw the entire circuit as professionally as possible freehand. You must make this first copy appear as near-perfect as possible. Once it is completed, it must be closely examined for any errors. Make certain that all components are schematically arranged in a logical order, with simplicity and ease of understanding as prime requisites.

Once you are satisfied that your freehand diagram is the best that can be produced, the templates may be brought into play and the diagram redone. Your first reproduction will most likely look far less professional. It will be necessary to practice for many hours before your drawings and tracings approach any level of professionalism.

It is assumed that you have reached a fairly accomplished state in drawing schematic diagrams freehand before moving on to the template method. However, it will be necessary to start from scratch, in a manner of speaking, to perfect this new style. When you start using templates, go back to the simple electronic circuits which you used as guides when you were learning freehand schematic drawing. A one-transistor circuit with a few resistors and capacitors is ideal. Again, use an example from a projects book or magazine. Choose only professionally reproduced schematics. Some projects books and magazines may present freehand drawings, which will serve no purpose here.

Try to make your template reproductions match those drawings that you find in these sources. This can be a rather disgusting phase of learning, but if you stick with it, the proper methods will come with time. Plan to spend many days or even weeks with a few simple schematic drawings. Don't be discouraged, because once you pass this stage, the drawing of moderate to complex circuits will come very quickly due to your past knowledge and experience. Moving up to complex drawings using the template method is simply a matter of reproducing a number of simple circuits and combining them accordingly. This statement should sound familiar, because it is the basis of all schematic drawing and reading.

Figure 5-10 shows a freehand drawing of a simple electronic circuit. Note that all components are neatly laid out in very logical positions to cut down on the lengths of conductor lines and condutor line crossovers. This is an ideal freehand drawing from which to make a professional reproduction. However, this drawing might be the end result of several other freehand drawings which were not nearly so neat or well laid out. Figure 5-11 shows what the first attempt at producing this freehand drawing may have looked like.

Obviously, this drawing has many flaws as far as representation is concerned. The schematic is electronically

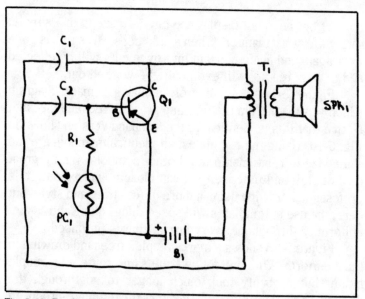

Fig. 5-10. Freehand drawing of a simple electronic circuit.

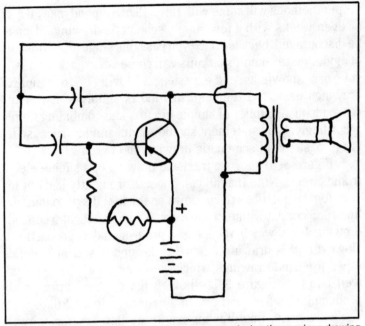

Fig. 5-11. An example of what a first attempt at producing the previous drawing might look like.

correct but does not display the circuit makeup in a simple, straightforward manner. When a drawing such as this is gone over again and again, certain improvements will be effected and a "best result" will eventually be worked out.

Figure 5-12 shows a professional schematic diagram which was made from the final and proofed freehand mockup. Notice that this is identical to the freehand version shown in Fig. 5-10. Using the templates, all identical symbols are the same size in relationship to all other symbols. With a great deal of skill and practice, you can closely approximate this professional drawing using a more or less freehand style, but here, the use of templates will speed things along and assure uniformity throughout every circuit that is produced.

Figure 5-13 shows a more complex freehand drawing in finished form. This one follows the same route as a simple circuit but probably took much longer to go through the debugging routine. Figure 5-14 shows the template-produced version of this same circuit.

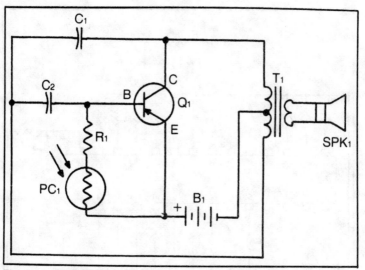

Fig. 5-12. Professional schematic diagram made from the final freehand mockup.

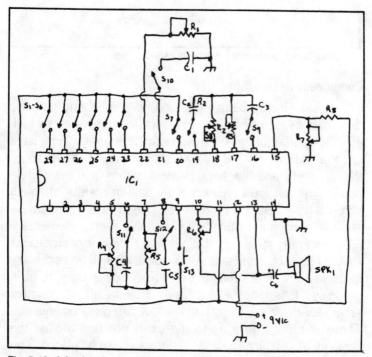

Fig. 5-13. A freehand schematic version of a complex circuit.

99

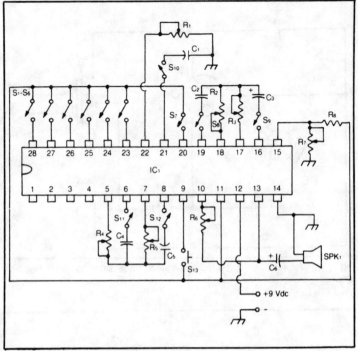

Fig. 5-14. Template-produced version of previous circuit.

Component Designations

It has already been explained that alphabetic/numeric designations are often given to electronic components which are displayed symbolically in a schematic diagram. Templates are available which will allow you to professionally produce these letters and numbers in many different sizes. Alternately, you can use a typewriter for similar results, although you will have less control over the size of each figure unless you use a machine which has switchable typing elements.

Generally speaking, all alphabetic/numeric designations should be of the same size. Occasionally it will be necessary to make a few alterations in order to get everything to fit in properly. This may be especially true in highly complex schematic drawings using hundreds of different components. There is no set form as to the exact position around the component at which these designators are to be placed. The rule here is to fit them in where you can and to make certain

100

that there is no question as to the component a particular designation identifies. Some draftsmen prefer to put the dessignator directly over all horizontally drawn symbols and to the left or right of those which are drawn vertically. Figure 5-15 shows a simple circuit composed of a transistor and a few other components. Note that the transistor is identified as Q1 and that this is placed directly overhead. Some resistors, on the other hand, are drawn vertically (in relationship to the way the schematic is normally read), and here, the R designations are to one side. There is standardization in the way the various electronic components are identified by schematic symbols and in the way these symbols are alphabetically/numerically referenced to a components chart.

To repeat, the numbers used with each letter will depend upon the number of any one type of component used in a schematic drawing. If there are ten resistors, for example, each will be given a designation of R and a different number from 1 to 10, in this case. If there were a hundred resistors, numbering would run from 1 to 100.

The alphabetic/numeric designations reference components to a separate list which provides the person reading the diagram with an exact description and/or part number for each component. Figure 5-16 shows a schematic diagram

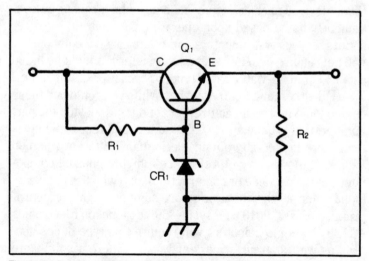

Fig. 5-15. Simple circuit showing alphabetic/numeric designations.

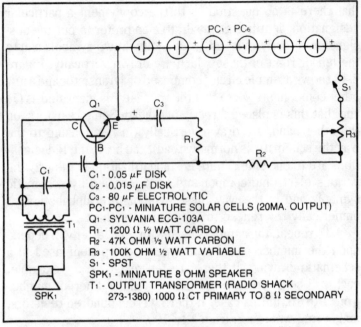

C1 - 0.05 μF DISK
C2 - 0.015 μF DISK
C3 - 80 μF ELECTROLYTIC
PC1-PC1 - MINIATURE SOLAR CELLS (20MA. OUTPUT)
Q1 - SYLVANIA ECG-103A
R1 - 1200 Ω ½ WATT CARBON
R2 - 47K OHM ½ WATT CARBON
R3 - 100K OHM ½ WATT VARIABLE
S1 - SPST
SPK1 - MINIATURE 8 OHM SPEAKER
T1 - OUTPUT TRANSFORMER (RADIO SHACK
273-1380) 1000 Ω CT PRIMARY TO 8 Ω SECONDARY

Fig. 5-16. A schematic diagram with matching components list.

along with a components list. Notice that this is arranged in alphabetical order, keying upon the alphabetic designators. Whenever possible, it is a good idea to give part numbers for those devices such as semiconductors which cannot be conveniently listed based upon operating parameters. Most resistors, capacitors, inductors and other passive components will not require this. For example, resistor R1 might be listed as a 120-ohm, ½-watt carbon type, or capacitor C1 might be a .01 μF disc ceramic-1000Vdc type. While you could find these components in a manufacturer's catalog and include the part number, this is unnecessary. Capacitors and resistors in particular are universally obtainable in many different common values. If, however, you reference a highly unusual component (due to its value or construction), one which is not easily found, then a manufacturer's part number might be appropriate, such as ".0013 μF Mylar—59 Vdc Johnson Electronics #112019." Some inductors will require this type of designation, as well as transformers and many different solid-state devices.

For this reason, it's a good idea for a person who is heavily involved in drawing schematic diagrams for use by others to obtain a semiconductor cross reference catalog from all of the major manufacturers (Sylvania, RCA, Motorola, General Electric, etc.). Unlike some other electronic components, all semiconductors are given a special designation which will differ from manufacturer to manufacturer. For example, a Sylvania transistor which is designated as an ECG 109 is equivalent to an RCA transistor with the designation of SK 3087. With a list of cross reference charts, you can easily modify older schematic diagrams to make them reflect the designations of today. A decade or so ago, many transistors and other solid-state components were given universal designation, sometimes referred to as generic designations. Examples would be 2N2139, 2N2222, and 2N3219. In those days, the 2N designation indicated that the device was a transistor, and the following four numbers indicated a specific type. Most companies which sold solid-state devices used these designations. Today, however, if you look at your local hobby store for a 2N2222, you probably won't find it. An SK3122 (made by RCA) is an exact replacement for the 2N2222, so the modern designation might best be used. This is not a terribly important point, as most people who build electronic circuits from schematic diagrams are accustomed to seeing many, many different types of semiconductor designations. Most of these people own a set of cross reference books and will simply reference to the company to which they have best access.

Manufacturer's cross referencing guides are useful from a design application as well, as they contain operating parameters for many of their devices. Some may even contain schematic diagrams of basic circuits used with these components. These can be a great aid to designing experimental electronic circuits schematically. It is often a good schematic diagramming practice to reference all solid-state components using a single company's designations. Choose any company you want, because the idea here is to allow the builder to use one catalog to cross reference all components (if necessary), rather than jumping from one manufacturer's catalog to another (if several different semiconductors are used in a single schematic diagram).

There are many other ins and outs in displaying electronic circuits schematically. Sometimes the draftsman will insert a note or two at the bottom of the drawing to indicate an unusual situation or a special part. Asterisks may often be used to reference components that may require modification to work within a given circuit. Good schematic drawing practice dictates that as much information as possible be relayed by means of schematic symbols. Sometimes, however, symbols are just not adequate enough to describe all that must take place, so the draftsman must revert to standard English. There is certainly nothing wrong with this, but do not get into the habit of using English descriptions in order to avoid a little more detail in the actual schematic. Plain language descriptions should be used only in a situation where a symbol will not suffice and where comprehension will be improved.

Summary

There is nothing especially difficult about learning to draw schematic diagrams. However, a tremendous amount of practice is required to become really proficient at this art. This chapter has laid out the basics of learning to draw schematics and has taken the reader through some stages of the professional side of this pursuit. Schematic drawing is a slow process, one that requires tremendous attention to detail. If you hurry, your learning proficiency will be impaired and, of course, the quality of your schematic diagrams will reflect this.

Students tend to rush through the early phase of training, hoping to quickly move into the realm of complex circuit design. Rush as they might, they get nowhere fast. If you play golf, you know that it is basically a simple sport. Anyone can be taught in a few hours how to swing a club and hit the ball. However, this is only a very small part of the entire game. Continuous practice is required in order to hit the ball *properly* and many years ensue before a professional degree of proficiency is obtained, if this comes at all. Looking at the full picture, golf is one of the most complex sports enjoyed today.

Schematic diagramming, fortunately, is not this complex, but there is far more to it than first meets the eye. Like golf, the only way to become proficient at schematic diagramming is through diligent practice. You will find, though, that it won't take years to reach a professional level.

Block Diagrams

It has been learned that complex electronic circuits are simply a number of simple circuits that have been combined. This has been a basis for our discussions regarding complex schematic diagrams and how to read and write them. Acting upon this same principle, another method of diagramming is sometimes used to represent electronic circuits. These are called block diagrams and are often used in conjunction with schematic diagrams to aid circuit comprehension and to speed troubleshooting procedures.

An earlier chapter dealt with a simple AM transmitter. This is reproduced in Fig. 6-1. This circuit was divided into six different sections, each of which performed a different electronic function. Figure 6-2 shows the block diagram equivalent of this circuit, which excludes all schematic symbols and represents each circuit section as a block. Additionally, each block is labeled with a description of the circuit it represents; i.e., mike preamplifier, audio amplifier, matching network, etc.

Block diagrams are quite useful to begin to explain how an electronic circuit is constructed. By starting first with the block diagram, we see the six major circuit sections. From a training standpoint, the microphone preamplifier section might then be displayed schematically and its operation

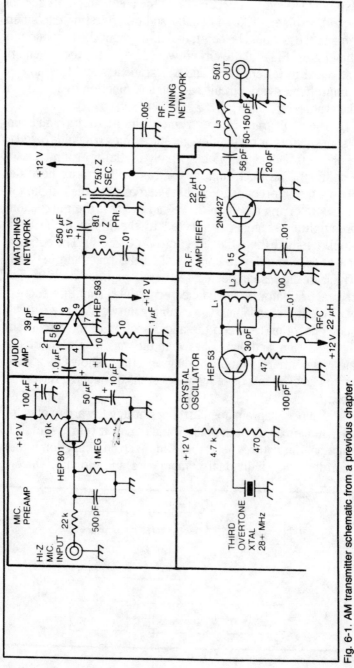

Fig. 6-1. AM transmitter schematic from a previous chapter.

107

explained. Next, the schematic of the audio amplifier section would be displayed individually and its operation explained. This process would be continued throughout the remainder of the circuit. Finally, the circuit would be presented schematically and in its entirety, and the student should be able to comprehend every circuit section and function by looking at the whole.

Block diagrams are extremely simple to draw and consist of squares or rectangles and sometimes triangles (used to represent a circuit block which is built around an IC amplifier). The block diagram used as an illustration here also shows the interconnections between each of the simple circuits. Referring to Fig. 6-2, we can see that the microphone preamplifier stage is connected to the input of the audio amplifier stage (note the direction of the arrow). The output of the audio amplifier is connected to the matching network, which in turn is connected to the rf amplifier section. The crystal oscillator is also connected to the rf amplifier section, whose output leads into the rf tuning network. There is only one connection between the audio section of the circuit and the rf section, and that is between the matching network and the rf amplifier. This block diagram, then, shows us a basic sequence of events or sequence of paths throughout the entire circuit.

A block diagram does little or nothing toward explaining the actual makeup of the electronic circuit it represents. It is functional in nature, describing an electronic function rather than depicting individual components and circuits. Once a

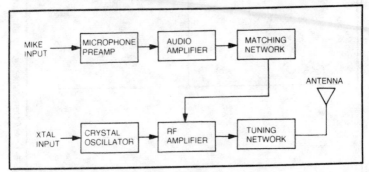

Fig. 6-2. Block diagram equivalent of Fig. 6-1.

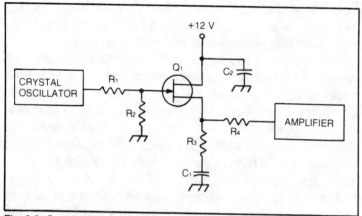

Fig. 6-3. Combination block/schematic diagram.

basic understanding of the principal circuit functions has been relayed by block diagram, then the schematic can be consulted for more practical details regarding troubleshooting or actual building.

Schematic/Block Diagram Combinations

Sometimes a block diagram and a schematic diagram are combined. This method is shown in Fig. 6-3 and is used when a particular circuit is to be highlighted and explained, especially as to its relationship to other circuits. The figure shows a buffer amplifier which is used in a radio frequency transmitter. A full schematic diagram is provided for the buffer circuit alone, whereas block diagrams indicate its relative position in regard to the master oscillator and the amplifier. This serves two purposes.

First of all, the person reading the schematic portion of the diagram can study the actual component makeup of the buffer circuit. He also is informed as to this specific circuit's place in the overall device. The block/schematic representation here indicates that the buffer receives its input from the master oscillator and channels its output to the amplifier. Another schematic diagram and block diagram combination might be used to describe another portion of this same device. Figure 6-4 shows an example whereby the master oscillator has been highlighted and indicates that the circuit channels its output to the buffer, which then inputs the amplifier. The only new information which is obtained here is contained in the schematic representation of the master oscillator.

Block diagrams may be handled in many different manners. Sometimes they are used to indicate interconnections between various pieces of equipment. When drawn as shown previously in this discussion, they are often called functional diagrams because they indicate the basic functioning of the electronic circuit. Figure 6-5 shows a functional diagram of a complex piece of biomedical equipment. To display the entire schematic diagram, many pages of this book would be required. The functional diagram, however, allows for an easy explanation as to how the device operates and will lead into further explanations provided by a schematic diagram of the complete system.

There are several ways block diagrams may be used. First of all, someone who is trying to arrive at a schematic diagram for a complex electronic circuit which must be designed more or less from scratch may decide to start with a block diagram. This would show in block form all of the circuit sections needed to arrive at a functioning device. This would be a functional block diagram showing how all circuit sections are interrelated. Then the designer would seek out schematic diagrams of circuits which could fill each block. In most instances, these circuits will have to be modified to blend with the overall system. The first block in the diagram would then be substituted by the schematic diagram of the circuit it specified. The schematic designer would move through the

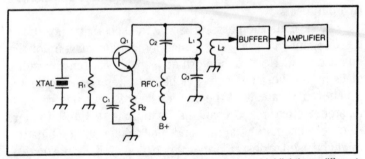

Fig. 6-4. Another example of a block/schematic diagram highlighting a different circuit portion.

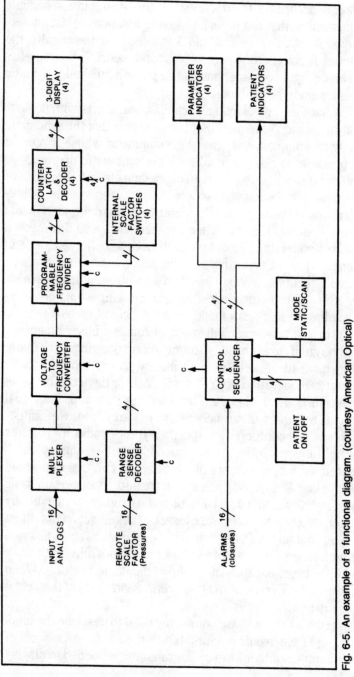

Fig. 6-5. An example of a functional diagram. (courtesy American Optical)

blocks according to functional order, designing schematic diagrams which can be used to build functioning circuit sections. As soon as the final block is filled in schematically, the device is complete. Eventually, a total theoretical circuit has been designed on paper and can be put in a finished schematic form, removing all blocks.

Another way of using block diagrams starts with possessing a finished schematic diagram. Assume that the schematic is very complex and that the equipment whose circuit it represents is malfunctioning. While schematic diagrams do describe the functioning of an electronic circuit, they are not as clear and basic as a functional block diagram. The technician would then laboriously identify each circuit section and draw it in block form. When finished, this would provide a clearer understanding of how each circuit section operates in conjunction with all others. Using this method, one or more circuit sections may be identified as a possible trouble area. At this point, the original schematic drawing would be referenced again and tests made in the indicated areas.

In practice, you will often encounter block diagrams, although if presented without accompanying schematic drawings, these will be used mainly to describe the basic functional operation of a class of circuit or device. The block diagram is used where a literal interpretation of individual circuit functions is not necessary. For example, we can describe the operation of a specific type of radio transmitter (amplitude modulated, continuous wave, frequency modulated, etc.) by means of a block diagram. This diagram would be applicable to nearly all radio transmitters of certain basic design. Now, no two types of radio transmitters built by different manufacturers are exactly alike, but all of them would contain the same basic circuit sections as far as function is concerned. One type of oscillator may work differently from another type, but they all perform the same function. When individual differences must be represented, then the actual schematic drawing is used.

Block diagrams are commonly used to describe the functioning of electronic circuits, but in the electronics world of computers, another form of diagramming is needed to display the functioning of a program. This is called *flowcharting* and is similar to block diagram representation, except the symbology is applied to the different basic sections of the computer program.

Flowcharts

The flowchart is a highly useful tool in computer programming and is a graphical representation of the paths which a computer program will take. Flowcharts are often prepared in conjunction with the generation of specifications and are modified as the requirements change to fit within the particular constraints of the overall system.

For complex problems, a formal written specification may be necessary to ensure that everyone involved understands and agrees on what the problem is and what the results of the program should be. To illustrate this, let's assume that a teacher wants a program that will determine a student's grade by calculating an average grade from six grades the student received over a grading period. The teacher will supply the six grades to the program as input. Only the average grade is needed as an output. Now, we can make an orderly list of what the program has to do:

1. Input the individual grades.
2. Add the grade values together to find their sum.
3. Divide the sum by the number of grades to find the average grade.
4. Print out the average grade.

We can also prepare a flowchart of the program, as shown in Fig. 6-6. As can be seen, a flowchart graphically presents the details of the structure of a complex program so that the relationship between parts can be easily understood. When the flow of control is complicated by many different paths that result from many decisions, a good flowchart can help the programmer sort things out. The flowchart is often useful as a thinking-out tool to understand the problem and to aid in program design. At this stage, then, the flowchart symbols should have English narrative descriptions rather than pro-

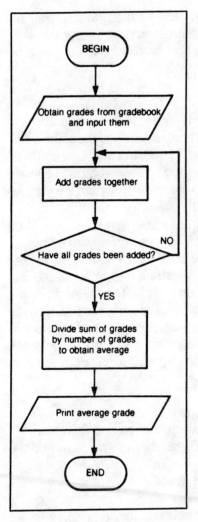

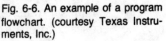

Fig. 6-6. An example of a program flowchart. (courtesy Texas Instruments, Inc.)

gramming language statements, since we want to describe what is to be done, not how it is to be accomplished. At a later stage, if formal flowcharts are required for documentation, the flowchart may contain statements in a program language. These flowcharts can be most helpful to another person who at some future time may need to understand the program.

Preparing a formal flowchart is time-consuming, and modifying a flowchart to incorporate changes is often quite difficult. Because of this, some programmers may express their dislike for this tool, but most will still use them as a development tool due to their invaluable assistance in understanding a program.

In order to promote uniformity in flowcharts, standard symbols have been adopted by several organizations. The symbols of the United States of America Standards Institute (USASI) are widely accepted, and some of their most commonly used symbols are shown and defined in Fig. 6-7. The normal direction of flow in a flowchart is from top to bottom

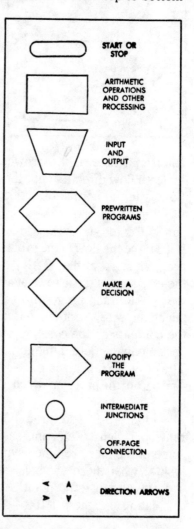

Fig. 6-7. Commonly used symbols of the USASI.

115

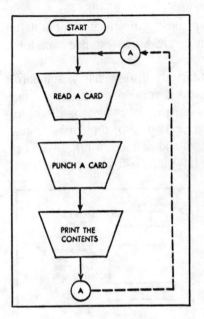

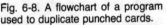

Fig. 6-8. A flowchart of a program used to duplicate punched cards.

and from left to right. Arrowheads on flow lines are used to indicate flow direction, but they are sometimes left off if the flow is in the normal direction.

Figure 6-8 shows the flowchart of a program to duplicate punched cards and at the same time print what is on the card. Let's trace the flow of the program through the flowchart. The program starts at the START oval at the top and proceeds in the direction of the arrows at all times. In the first box below the START, the program reads a card. The next step is to punch the card's contents into a blank card and then print out the contents on the console printer. The program then goes back along the dotted line to the top and reads the next card. The program repeats itself as long as there are cards to read. The part of the program which is done over and over is called a *loop*.

Figure 6-8 used three different symbols to mean different operations. Referring back to Fig. 6-7, which showed some of the more common symbols used in flowcharts, let's look at what these symbols mean. Oval boxes are used to show a start or stop point. Arithmetic operations such as calculating "height $= 1,100 - 16t^2$" are placed in rectangular boxes. Input and output instructions would be placed in an upside-down trapezoid. If we used a program written earlier in a program, we would not bother drawing the flowchart for the inside program. Rather, we would just place the entire program inside the flattened hexagon. If a box is to be used for making a decision, the diamond shape is used. A five-sided box is used to show a part of the program that changes itself. A small circle identifies a junction point of the program. This is a point in the program that is connected to several places, and we use the intermediate junction symbol to avoid drawing long lines on the flowchart. A small five-sided box is used to show where one page of a flowchart connects to the next (if more than one page is used for the same flowchart). The intermediate junction and off-page connection would further be labeled with a number or letter, so that all like symbols with the same letter or number inside are connected together. Finally, arrows show the direction of travel.

Returning to the flowchart for duplicating punched cards, let's suppose that we want to change the flowchart so that the computer skips any blank cards and duplicates only those cards with some information punched in them. Since the computer must now make a decision about each card, the decision block will be needed in the flowchart, which will be changed to appear as that shown in Fig. 6-9.

Except for the decision block, the flowchart shown here is the same as the earlier one. We again begin in the START oval at the top and then go to the block marked READ A CARD. But now the program will go to the decision block labeled CARD BLANK? If the answer is yes, we go left to the connection circle A and back to the top to read the next card. Only if the card is not blank do we go right and actually punch a duplicate card and print its contents.

The flowchart we have used as an illustration here is a very simple one using only input and output devices and doing no calculations. Most programs and flowcharts are not this simple, and this one is used as an example to present the reader with an idea as to how a flowchart graphically portrays what a computer program actually does.

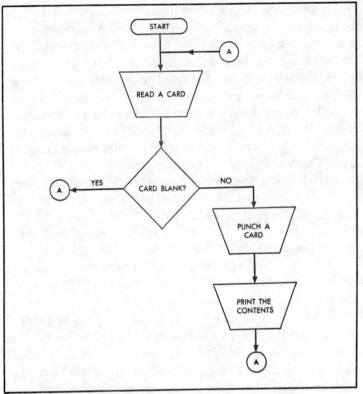

Fig. 6-9. Example of a flowchart which includes a decision block.

The field of microcomputers uses a great deal of different types of diagrams. There are flowcharts, Venn diagrams, etc. These deal mostly with the software portion of this industry. From a purely electronic standpoint, functional diagrams abound and are usually more numerous than the schematic diagrams. From an understanding standpoint, block diagrams are adequate to display all machine functions, but repair necessitates well-defined schematic drawings. Computers take advantage of the latest state-of-the-art developments in electronic components and are relatively simple from this standpoint, especially when you consider all they can do. However, from a pure electronics standpoint and as far as schematic diagrams are concerned, they are highly complex and many pages of schematics are required to represent even the simpler units.

Summary

Block diagramming is an excellent basic form of understanding the functioning of electronic circuits. This type of diagramming is very easy to do from a drafting standpoint and usually requires only a straight edge. Most block diagrams can be drawn in a relatively short period of time when compared with their schematic equivalents. Before these can be produced, however, it is usually necessary to have a good understanding of schematic drawings and what they represent as far as circuit construction and circuit function is concerned. The complex part must be done before this simple method of function description can be presented in a logical and accurate manner.

Pictorial Diagrams

Throughout this book, it has been stressed that symbology is the key to understanding how to read and draw schematic diagrams. Symbology is ideal for expressing the complex working of electronic circuits in a medium which can be readily accepted and processed by the human brain. Sometimes, however, we can use another type of diagram to aid in circuit understanding. At several points in this book, pictorial diagrams have been used to show the components which our schematic symbols represent. Pictorial diagrams are often used in conjunction with schematic drawings to indicate physical relationships. Remember, schematic diagrams don't really do this. Rather, they show electrical and electronic relationships. Theoretically, an electronic circuit which can be built on a piece of circuit board one inch square could also be built on a circuit board one mile square. The components would simply be spread out and great lengths of hookup wiring would be used to interconnect them. From a practical standpoint, the resistance losses that would be incurred here would not allow the circuit to operate, but from a theoretical standpoint, physical relationships (component placement) are not usually dealt with.

When building or servicing some electronic circuits, the physical relationship of one component to another plays a very important role. Conductors will have certain inductive and capacitive effects and will act directly upon the circuits. The actual effect will depend upon the frequency of operation and many other factors. If two components are to be positioned in very close proximity to each other, the schematic cannot show this in a practical manner. Certainly, an English language notation can be made on the schematic drawing, but a more efficient method involves the use of a pictorial diagram to show the actual components in the circuit.

Pictorial diagrams may be two dimensional in appearance or three dimensional. Some may be drawn to look exactly like the finished circuit. Others may be drawn in a quasi-block diagram form which roughly indicates the physical size of components along with physical spacing.

Figure 7-1 shows a schematic diagram from an electronics projects book I wrote some years ago. This diagram provides all of the electronics relationships needed to build a working model but tells the builder nothing about how the components are to be physically situated. This particular circuit is best built on a small piece of perforated circuit board, so the drawing shown in Fig. 7-2 was also included to allow the builder to see how this latter job is to be accomplished. Notice that the component references are also included so that proper cross referencing can be done between the two diagrams.

This is a pictorial diagram and it can be classified as basically a two-dimensional one. Height and width are indi-

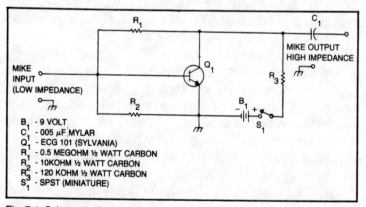

Fig. 7-1. Schematic diagram of a simple circuit from an electronics project book.

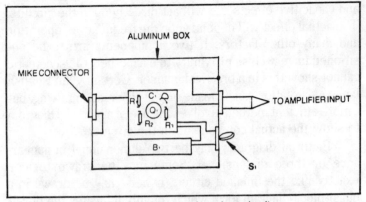

ALUMINUM BOX

MIKE CONNECTOR

TO AMPLIFIER INPUT

R₁   C₁   Q₁

R₂   R₁

B₁

S₁

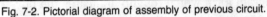

Fig. 7-2. Pictorial diagram of assembly of previous circuit.

cated in the drawing, but depth is not. The circuit board is in the exact center of a box which indicates the aluminum compartment the overall circuit is to be mounted in. The physical placement of the mike connector, the switch, the circuit board, the battery, and the output plug are all indicated. This one drawing allowed the builder to see how the five components should be situated on the circuit board and how the other components could be mounted to the box. Now, the builder really has something to work from—a schematic diagram *and* a simple pictorial drawing.

While the above example is a pictorial drawing, the components shown do not look exactly like they will in actuality. This is still a language of symbology, but the symbols are much more closely related to the general physical appearance of each component. The builder now has an idea of what his circuit should look like once it is completed.

Electronic components are usually very simple to draw in this manner. A pictorial of this type shows an overhead view, which is necessary when only two dimensions are to be indicated. A side view of this circuit would require that height (or length), width *and* depth be displayed. When I speak here of three dimensions, I should also say that it is impossible to truly display three dimensions on paper, because, of course, line drawings are always two-dimensional. We can, however, simulate three dimensions on paper. This will be dealt with later.

Recapping just a bit, it is again necessary to point out that a schematic diagram gives us the electrical and/or electronic relationship of the components in the circuit, while the pictorial drawing shows us the physical relationship. Look at the last figure again. It tells us nearly nothing about the electronic relationship of the components in the circuit. The only electronic relationship that can be shown involves the battery (B1) and switch (S1). We can see here that one lead of the battery is connected to one switch contact. The remaining battery lead and switch contact lead enter the circuit board at this point. Looking back at the schematic diagram of the same circuit, it can be seen that the positive battery lead does indeed connect to one of the switch terminals. The remaining switch terminal is connected to resistor R3. In the pictorial drawing, however, R3 is part of the circuit board and the physical relationship stops pictorially before this component is encountered.

If you had experience at building simple electronic projects, then the pictorial drawing would not be necessary in this case. For those who do not have many hours of bench experience, however, the circuit might never have been built with the schematic diagram alone. From a service standpoint, suppose this circuit were built and then failed suddenly. Assume also that a single resistor had burned up due to an overload and its value could not be identified. By referring to the pictorial diagram, you could quickly locate the burned up component and then reference it to the schematic diagram, which would in turn reference it to the components list. A new component could then be purchased and wired back into the circuit using the two types of diagrams.

How do you draw pictorial diagrams? Obviously, it is necessary to have a finished circuit on hand or to be experienced enough to accurately visualize it in your mind. Assuming that you have a circuit on hand, all you do is visually examine it and then start building your pictorial diagram, one component at a time. This is the same method used to read and draw schematic diagrams. If the entire circuit is to be mounted on a piece of circuit board, then start by drawing a likeness of the board on paper. If the original board is ten

inches by five inches, it may be necessary to reduce it some-
what. A representation that is five inches by two and a half
inches will be exactly half the size of the original board.
Therefore, all electronic components should be drawn half-
size as well. In practice, the exact dimensions are relatively
unimportant. Simple circuits don't usually require strict
adherence to physical placement which is accurate down to a
fraction of an inch.

A simple pictorial diagram is shown in Fig. 7-3. This was
taken from a book written by Isaac R. Holstroemn entitled
Energy From The Sun—33 Easy Solar Projects (TAB Book
1323). Here, the circuit board is indicated as a rectangle and
five components are placed on it at various points. The circuit
layout is a general one showing the transformer at the left-
hand side of the board and the components arranged slightly
right of center. The transformer is shown near the edge of the
board because its output leads will feed a speaker. If the
transformer were in the center of the board, an extra length of
wiring would be required to access the speaker. The purpose
of most pictorial drawings of electronic circuits is to indicate
how the components may be placed physically to limit the
interconnecting wiring. Extra wire means extra circuit resis-

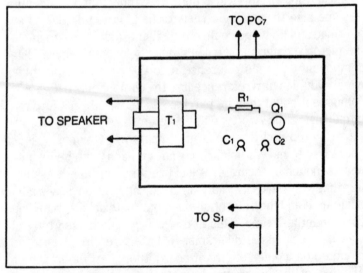

Fig. 7-3. A pictorial diagram of a simple electronic circuit.

tance, as well as increase inductance and capacitance. The schematic diagram for the pictorial drawing discussed here indicated that C1 and C2 were connected together at one point. C1 and C2 are shown in close physical proximity due to this. If C1 were moved to the far left-hand side of the board, then an extra length of wire would have to connect the two. The extra resistance could affect the circuit's operation. Pictorial diagrams, then, are usually based on good electronic building practice, which stipulates that component lead wiring be kept to an absolute minimum.

The examples of pictorial diagrams shown are what I classify as *pictorial block drawings*. They do show the physical relationships and more closely symbolize the electronic components, but they do not show a true-to-life picture of the finished circuit.

When it becomes necessary to use a pictorial drawing to locate specific electronic components, a presentation similar to the one shown in Fig. 7-4 may be used. This is a reprint from a Hewlett Packard service manual showing a circuit board from a fetal monitor. This is a highly complex piece of equipment, and each circuit board may contain hundreds of different components. Trying to locate one of these on a circuit board can be very difficult, so the board is drawn actual size and an identification grid is placed around it. In this example, the circuit board is sectioned off from 1 to 8 vertically and in two sections (A and B) horizontally. A chart is included which shows the component's designation and its grid location. Locating C1 (at the top of the chart) involves examining the area in the grid of B2. All you do is go to the right-hand section of the circuit board (the B grid) and then look in the area encompassed by the grid marked by the number 2. C1 can readily be identified, although it may take a bit of looking. Some schematic diagrams will also be handled in this manner.

Notice that in lieu of drawing the components on this particular example, some are indicated only by their alphabetic/numeric designation. But a line is drawn on both sides of this designation to indicate the length of the component and, in part, the space it occupies.

REF DESIG	GRID LOC	REF DESIG	GRID LOC	REF DESIG	GRID LOC	REF DESIG	GRID LOC
C1	B2	C45	A6	Q10		R34	A3
C2	B2	C46	A6	Q11		R35	A4
C3	B2	C47	A7	Q12	B5	R36	A4
C4	B2	C48	B7	Q13		R37	A4
C5	B3	C49	A7	Q14	A5	R38	B4
C6	B3	C50	A7	Q15	A5	R39	A4
C7	B3	C51	A7	Q16	B5	R40	A4
C8	B3	C52	A8	Q17		R41	
C9	B3	C53		Q18		R42	
C10	B3	C54		Q19	B8	R43	B3
C11	B4	C55	A4	Q20	A8	R44	A4
C12	B4	C56	A4	R1	B2	R45	A4
C13	B3	C57	B6	R2	B2	R46	A4
C14	B4	C58	B6	R3	B2	R47	
C15	A2	CR1	B3	R4	B3	R48	
C16	A2	CR2	B3	R5	B3	R49	B5
C17	A2	CR3	B4	R6	B4	R50	
C18	A2	CR4		R7	B4	R51	B5
C19	A2	CR5	A2	R8	B4	R52	B5
C20	A2	CR6	A2	R9	B4	R53	A5
C21	A3	CR7	A2	R10	B4	R54	A5
C22	A3	CR8	A2	R11	B4	R55	A5
C23	A3	CR9		R12		R56	A5
C24	B3	CR10	A2	R13		R57	B5
C25	A3	CR11	A4	R14	A2	R58	B5
C26	A3	CR12	A4	R15	A2	R59	B5
C27	A4	CR13	B8	R16	A2	R60	B5
C28	B6	CR14	A6	R17	A2	R61	B5
C29	B6	CR15	A6	R18	A2	R62	
C30		CR16	A7	R19	A3	R63	B6
C31	B5	L1	B2	R20	A2	R64	B6
C32	B5	L2	B2	R21	A2	R65	B6
C33	A5	L3	B3	R22	A2	R66	B6
C34	A5	L4	B3	R23	A4	R67	B6
C35	B6	L5	A3	R24	A3	R68	B6
C36	B6	Q1	B3	R25	A3	R69	B6
C37	B7	Q2	B4	R26	A3	R70	B6
C38	B7	Q3		R27	B2	R71	B6
C39	B7	Q4		R28	A2	R72	B7
C40	B7	Q5	B2	R29		R73	B7
C41	B7	Q6	A2	R30		R74	B7
C42	B7	Q7	A3	R31	B3	R75	B7
C43	B7	Q8	A3	R32	A3	R76	B7
C44	B7	Q9	A3	R33	A3	R77	B7

NOTE

Components C47, C48, R105 are used on Beat-to-Beat Board only

Components C39, C40, C44, C51 and R106 are used on Average Board only

Fig. 7-4. A pictorial diagram which includes a location grid (courtesy Hewlett-Packard).

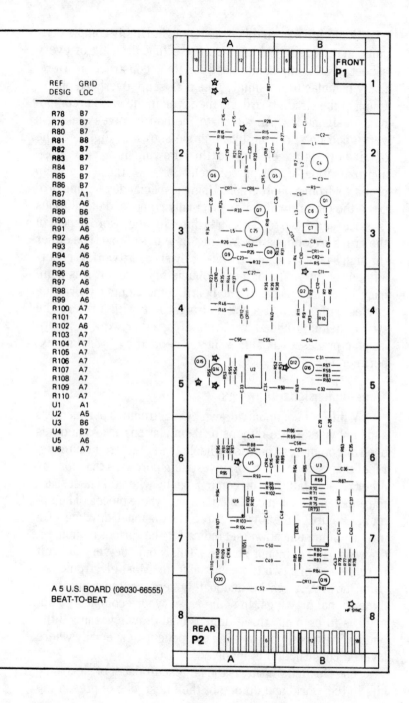

REF DESIG	GRID LOC
R78	B7
R79	B7
R80	B7
R81	B8
R82	B7
R83	B7
R84	B7
R85	B7
R86	B7
R87	A1
R88	A6
R89	B6
R90	B6
R91	A6
R92	A6
R93	A6
R94	A6
R95	A6
R96	A6
R97	A6
R98	A6
R99	A6
R100	A7
R101	A7
R102	A6
R103	A7
R104	A7
R105	A7
R106	A7
R107	A7
R108	A7
R109	A7
R110	A7
U1	A1
U2	A5
U3	B6
U4	B7
U5	A6
U6	A7

A 5 U.S. BOARD (08030-66555)
BEAT-TO-BEAT

Sometimes even this is not enough. Highly complex circuits may require a highly accurate line drawing of every aspect of all components used for their construction. Here, many manufacturers simply take a closeup, overhead photograph of the circuit board and then type in the various component designations. An alternate method involves making a very good line drawing like the one shown in Fig. 7-5. The detail here is excellent and this drawing is a very close approximation of the photograph. Some of these are even prepared from a photograph. Using a special developing process, the picture which was originally composed of halftones is processed down to black and white. In other words, most of the grays are eliminated. Making drawings of this type either through a photographic process or by sheer artwork is generally out of the realm of most hobbyists. However, it is quite easy to approximate circuit board components by drawing circles with a compass and a straight line with a ruler. The type of detail provided by extremely good artwork or photographic processes is almost never needed by the home experimenter.

Three-Dimensional Drawings

While not commonly drawn by electronic experimenters and hobbyists, three-dimensional simulations are sometimes used by commercial outfits to display their electronic circuits. I feel this is done more for advertising purposes than for any other reason, but they do come in handy when it is necessary to represent mechanical functions and whole pieces of equipment. Figure 7-6 shows a three-dimensional drawing of a military communications set indicating the various equipment drawers used to mount devices. This would be very difficult to represent in a two-dimensional form. Most electronic circuits, however, can be adequately represented using two-dimensional drawings. In some cases where complex equipment is to be built, three-dimensional drawings can aid the builder in properly positioning components, especially where layered construction is involved.

The Heath Company is the largest producer of electronic kits in the world and do one of the finest jobs of presenting

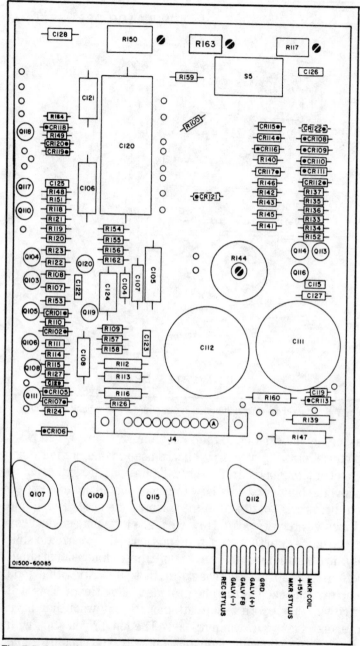

Fig. 7-5. An almost true to life line drawing of an electronic circuit (courtesy Hewlett-Packard).

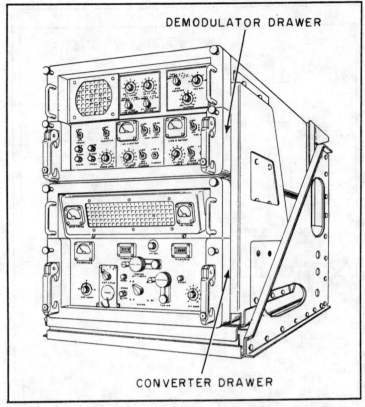

DEMODULATOR DRAWER

CONVERTER DRAWER

Fig. 7-6. Three-dimensional line drawing of a military communications set.

assembly information to their customers. Figure 7-7 shows a pictorial drawing of a circuit board used in one of their electronic microphones. Here, three dimensions are simulated to show the builder exactly how the components are to be fitted on the circuit board and in the microphone case. These line drawings are very good. However, this type may be made by electronics experimenters who possess a bit of artistic ability. In order to learn to draw simple three-dimensional simulations, it is absolutely mandatory that you start with a good line drawing such as the one provided us by Heath. This will give you the opportunity to attempt to copy what they have presented for your own purposes. The longer you stick with it, the better you will become. Eventually, you should be able to do a fair job. This book has taken three-dimensional draw-

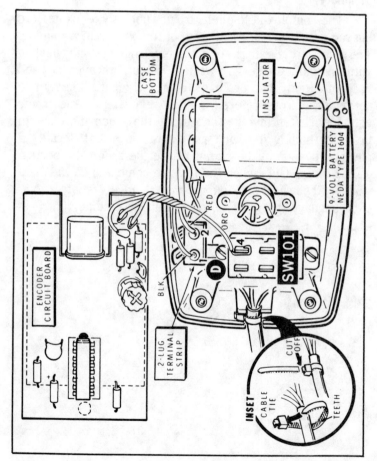

Fig. 7-7. Three-dimensional diagram of a heathkit electronic microphone. (courtesy Heath Company)

ings into account to allow the reader to know how and why they are used. However, we are getting into a totally different realm which does not lend itself readily to the beginner. A person who has had no experience whatsoever with schematic diagrams can quickly learn to read and draw them. Pictorial diagrams can be read even more quickly, but drawing them involves the development of artistic ability which is far beyond the scope of this book. If you intend to make electronic diagramming your profession, then you are urged to take some basic art courses at a local college. Courses in electronics drafting will also be of a great deal of assistance.

Summary

The complex field of electronics not only lends itself to but requires many different types of symbology diagramming. The pictorial diagram is a highly useful language when dealing with assembly and service work. It cannot, however, be used very efficiently without a matching schematic diagram to serve as a cross reference. The servicing of electronic equipment often involves consulting the schematic diagram first, then pulling in the pictorial diagram in order to locate the physical portion of the circuit where the schematic portion can be found. Only when the two are combined do the electronic and physical relationship align themselves into a complex language that aids the overall understanding.

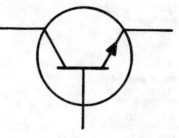

Index

133

Edited By Steven Mesner